Flora

of the

Christchurch Area

Felicity Woodhead

Published by Felicity Woodhead, 28 Hungerford Road, Bournemouth, BH8 0EH.

ISBN 0 9522857 0 3

Typeset by Greenlink, Bournemouth, Dorset.

Printed in Great Britain by Blackmore Press, Shaftesbury, Dorset.

Front cover: Christchurch Priory from Stanpit Marsh.
Inset: *Orchis morio* Green-winged Orchid.
Original painting by Jonathan Tyler

Back cover: *Gentiana pneumonanthe* Marsh Gentian.
Photograph by Keith Woodhead

Sponsorship: Greenlink
The Friends of Stanpit Marsh
Botanical Society of the British Isles
English Nature

CONTENTS

LIST OF FIGURES

LIST OF COLOUR PLATES

To Terry who inspired me,
and to all those who
supported me through the difficult times.

FOREWORD

One of my happiest memories during the preparation of the <u>Atlas of the British Flora</u> in the middle 1950s is of spending a week with N.Douglas Simpson at his house 3 Cavendish Road, Bournemouth. Here he had gathered a unique collection of British local Floras, the largest private collection in the country, and together we abstracted the distribution records for a single potentially underworked county - Selkirk. I realised then, for the first time, just how extensive and important is the information that local Floras contain, in some areas like Cambridgeshire, going back over 300 years.

Nowhere else in the world can boast such a continuous study: no flora is better known than that of the British Isles. As I write this at the end of 1993 I reflect with astonishment that, since the publication of the <u>Atlas</u> in 1962, Floras have been produced for nearly every county in England and Wales - five in the last 6 months. Most of these have gathered information using a 2 x 2km 'tetrad' as their recording unit appreciating that this, which gives 500-1000 units in an average county, was as much as any individual or group could manage in a reasonable period like a decade.

But this refinement of detail compared with the 10km squares of the <u>Atlas</u> is still too coarse to reflect the subtleties of the local distribution of our wild flowers in relation to soils and topography and to man-made features like roads, railway lines, canals and quarries. For this a unit like the 1km square is ideal - but inevitably, because of the work involved, cannot be used in an area as large as a county.

So I warmly welcome the enterprise of Felicity Woodhead in adopting the 1km square and concentrating her work on a small area which has a wide range of habitats and where change has been rapid. It provides information which Floras using a larger unit cannot give. Her <u>Flora</u> will grow in value with time and future botanists, and conservationists, will be even more grateful to her than even we are, for having recorded single-handed, so assiduously and accurately, the flowers and ferns of her corner of England.

In doing so she has joined the ranks of the hundreds of dedicated field botanists before her who have given time to, but have gained enormous enjoyment from, the pursuit of plants around them. May the habit never die!

Franklyn Perring, Green Acre, Wood Lane, Oundle. December 1993

President, Botanical Society of the British Isles

ACKNOWLEDGEMENTS

I should like to thank all those people who have helped me, not only with the identification of specimens, but also those who have spent many hours accompanying me on my fieldwork, especially my husband and daughters, who have encouraged me and repeatedly returned to some of the same sites many times. For identifying most of the Downy Birch in the area, I thank my youngest daughter Alex, and for her help with typing I thank Jill.

In particular I would like to thank Martin Jenkinson, Paul Bowman, Dr Humphry Bowen, David Pearman, John Ounsted, Brian Edwards, the late Cecil Pepin, John Lavender, Stuart Clarke, and Richard Surry of DERC for their patient and understanding botanical help over the past ten years, and also all the referees who have checked the identification of species for me. I would also like to thank Dr Roger Cooper, Dr Alan Morton and my husband Keith for persuading me to computerise my records, and for supporting me through many perplexing hours of computing, and especially to Alan Morton for permitting me to use his DMAP computer program for producing the species maps.

I would like to express my special thanks to Jonathan Tyler for designing and drawing the cover, and to Ann Percy for producing the plant drawings and to Keith for taking most of the photographs. I am particularly grateful to Richard Hands and Joyce Knowles of Greenlink for their support and expertise in organising the desktop publishing, and Judith Plumley of Christchurch Borough Council for her support and encouragement. I would also like to thank the Friends of Stanpit Marsh for their donation towards the cost of the colour cover and the BSBI for their grant towards publication costs. My thanks also go to all those who are too numerous to mention individually, especially those who have given me information concerning the locations of particular species and sites. Without the help, encouragement and understanding of so many people this Flora would never have been produced.

Felicity Woodhead, Bournemouth 1993

LIST OF ABBREVIATIONS

ABF	Atlas of the British Flora, 1962
agg.	aggregate: includes two or more species closely resembling each other
BEE	Bournemouth Evening Echo
BNSS	Bournemouth Natural Science Society
BRC	Biological Records Centre, Monk's Wood
BSBI	Botanical Society of the British Isles
CTM	Clapham, Tutin and Moore, 1989
DERC	Dorset Environmental Records Centre
km	kilometre
NCC	Nature Conservancy Council now English Nature
PBNSS	Proceedings of the Bournemouth Natural Science Society
PDNHAS	Proceedings of the Dorset Natural History and Archaeological Society
pers.comm.	personal communication
sens.lat.	sensu lato - in the broad sense
sens.str.	sensu stricto - in the narrow sense

INTRODUCTION

One of the main reasons for writing this Flora is the pressing need for up-to-date information on the distribution of plant species in the rapidly changing environment of southeast Dorset. The last Flora concentrating on this area was written in 1900 by Rev. E.F. Linton, who studied an area within a twelve mile radius of Bournemouth Square. Some years before, in 1883, a Flora of Hampshire by F. Townsend was published, followed by a second edition in 1904. These also covered most of the area around Bournemouth and Christchurch, although in less detail than Linton's Flora. The Kinson area, however, was not included as it was then in Dorsetshire. Some additional records around Bournemouth were added in two appendices to Linton's Flora in 1919 and 1925, and in 1929 J.F. Rayner produced a supplement to Townsend's Flora of Hampshire.

The two main Dorset Floras were written by Mansel-Pleydell in 1874 with a second edition in 1895, and by Good in 1948 with a second edition in 1984. Neither of these included the area around Christchurch, which was then in Hampshire, although Mansel-Pleydell did cover the Kinson area; but by 1948 Kinson was part of Bournemouth Borough and in Hampshire, and was therefore not included by Good.

Apart from these Floras there is a relative poverty of published records between 1904 and 1980 for the area. The principal source of records is the Bournemouth Natural Science Society Proceedings, and in particular those records collected by L.B. Hall and N.D. Simpson in the 1920s.

The area covered by this Flora is not defined by a natural boundary, but by an artificial one. It consists of Christchurch Borough, the Parish of St Leonards, and parts of Bournemouth Borough including the cliffs, chines, the Stour valley and Kinson and Turbary Commons (see Figure 1). These are all former parts of Hampshire which were transferred to Dorset in the Local Government reorganisation of 1974. However, almost all of the area remains within the botanical vice-county 11, South Hampshire, with the exception of the area around Kinson and Wallisdown which is included in vice-county 9, Dorset (see Figure 2). This additional area of Dorset was mentioned, but not included, in Good's Concise Flora of Dorset (1984) as he stated that "most of it (the area) is heavily built up, so that the additions may not be many" (Good, 1984, p.iv) It will, however, be covered by the new Hampshire Flora due to be published shortly, although not on a 1 kilometre grid square basis.

Since 1900 when Linton and Townsend were studying the local flora there have been considerable changes to the countryside, especially around Bournemouth. The town itself has grown from a population of 59,760 in 1901 (General Register Office, 1901) to over 154,680 in 1991 (OPCS, 1991), and the resulting changes in the area covered by the town can be seen in Figures 3 and 4. Many of the locations which Linton gives are now substantially altered, and some of them have been totally destroyed; for example St Mary's Common, the heaths of West Cliff and Winton Heath are no longer large areas of heaths and commons as they were then.

However, even before 1900 development was causing loss of species. For example, in Linton only three locations for *Vicia lathyroides* are given; one on Mudeford Shore in 1879, another near Swanage and the third behind Christchurch Railway Station. Linton states "The Mudeford Station is built over; I could not find any in 1885," (Linton, 1900, p.75) although he later recorded some from the gravelly yard behind Christchurch Railway Station in 1893. Townsend in his preface to the second edition states:

> "It is sad to think that our native Flora is suffering much, even to the extinction of species, by building and enclosures in the neighbourhood of our larger towns, whereby the localities of many plants have been lost entirely. Marshes have also been extensively drained and much land laid out in pleasure-gardens, market-gardens, and for recreation purposes."

(op. cit.,p.v).

As well as many natives, some non-native species also were destroyed by housing developments before 1885. For example *Montia perfoliata* was lost from Mudeford even though it was abundant there in 1879, only six years previously (Townsend, 1904, p.75). As Townsend stated in his preface, many sites have suffered from drainage and other physical alterations, and species such as *Hypericum elodes* have decreased through the drying out of their habitats. Changes in agricultural practice have also led to the decline of particular species. The improved cleaning and sorting of arable seed for instance, has led to the decline of *Centaurea cyanus*, and the use of fertilizers and herbicides has contributed to a significant decline of several species including *Misopates orontium* and *Myosurus minimus* (Wilson, 1992).

More recently other habitats have been destroyed, although not necessarily by human action alone. For example, since the 1950s the extensive sand dunes between Mudeford and Friars Cliff have been lost due to a combination of beach development and coast erosion. In addition the nearby sand dunes at Hengistbury Head have been trampled unremittingly by the thousands of holidaymakers visiting each summer, and in consequence the flora has suffered, in some cases to extinction. *Eryngium maritimum* has not been recorded at either Mudeford or Hengistbury for many years now, although it was recorded between 1930 and 1962 (ABF, 1962).

Appendix 1 provides a list of native species which have been recorded historically in this area, but not found during the present survey. Over forty of these species must now be presumed to be extinct in the survey area.

In addition there are at least 120 species which have apparently declined considerably in abundance in this area since Linton's and Townsend's Floras were produced. These are described in the individual species notes. Not all species have declined though; there are over 30 species which seem to have increased in distribution since the early 1900s. For example, *Crepis vesicaria* has considerably increased, and *Juncus tenuis* and *Lactuca serriola* were not mentioned by Linton or Townsend, but are now frequent in some places. Unfortunately, several of the species which have increased are invasive, such as *Reynoutria japonica* and *Gaultheria shallon*.

In the period of almost 100 years since the last local Flora for this area appeared the changes are obviously many, and for this reason it is important to have up-to-date information on the current distribution of plant species in the area.

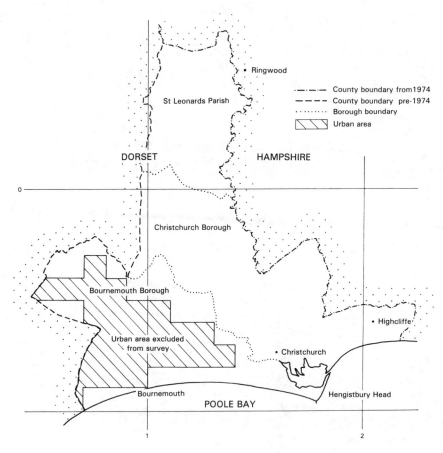

FIGURE 1. Area covered by the Flora

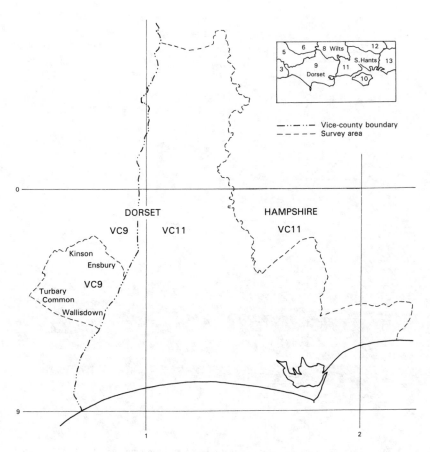

FIGURE 2. Vice-county boundaries

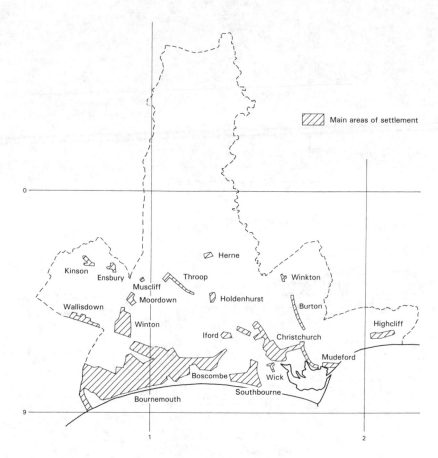

FIGURE 3. Christchurch and Bournemouth in 1901

FIGURE 4. Extent of the built-up area around Christchurch and Bournemouth in 1991

© Crown copyright

GEOLOGY, SOILS AND HABITATS

The area of southeast Dorset covered by this Flora lies within the Hampshire Basin and is composed of Tertiary rocks. These consist of the Bracklesham Group overlain in places by the Barton Group and River Terrace Deposits and other drift deposits (Bristow, Freshney & Penn, 1991). The surface geology, as shown in Figure 5, is fairly complex, and only a brief description will be given here. Figure 5 is based on 1:50,000 British Geological Survey Maps 329 (1991) and 314 (1976), by permission of the Director, British Geological Survey; NERC copyright reserved.

The Bracklesham Group comprises the Poole Formation and the Branksome Sand Formation. These are roughly equivalent to the former Bagshot Beds and the lower part of the Bracklesham Beds, previously called the Bournemouth Marine Beds. They consist mainly of sands with brickearth and clays. The beds of the Poole Formation outcrop only in small areas mainly around Kinson and also in some valleys in and around Bournemouth. The uppermost beds of the Poole Formation, the Parkstone Clays, also outcrop around the sides of St Catherine's Hill and at Ensbury Wood. The Branksome Sands outcrop at Parley and from St Catherine's Hill northwards between the Moors River and the River Avon, to St Leonards and Ashley Heath. There are also small areas of outcrop around Bournemouth, where the Branksome Sands are generally overlain by River Terrace Deposits. The soils in these areas are sandy and very acid, and now support mainly heathland and coniferous woodland habitats, although much of the higher land in the southwest coincides with the densely populated urban area of Bournemouth and is excluded from the Flora.

There are higher areas of River Terrace Deposits also in the extreme east of the area at Highcliffe, on Hengistbury Head and St Catherine's Hill, at Throop, and from Matchams View northwards to Ashley and the county boundary. These areas have well-drained sandy soils, sometimes with flints, and the main habitat is heathland.

The Barton Group now consists of the Boscombe Sand, Barton Clay, Chama Sand and Becton Sand Formations. The Boscombe Sand is composed of sands, clays and pebblebeds, and formerly comprised the upper parts of the Bracklesham Beds. The Boscombe Sand is exposed along the cliffs and there are a few small outcrops near Boscombe, from Southbourne to Littledown and at Highcliffe. The soils on the Boscombe Sands are generally very acid sandy soils sustaining mainly heathland and coniferous trees. Along the coast at Hengistbury Head and Highcliffe and in nearby valleys there are bands of Barton Clay exposed. On these clays many small strips of deciduous woodland still exist.

Three principal rivers flow through the study area (see Figure 6), and extensive River Terrace Deposits are present, in particular to the east of the River Avon, south of the River Stour and between the Stour and Moors River valleys. These areas have well-drained coarse loamy and sandy soils and arable farming and horticulture are more frequent here than elsewhere. The floodplains of the Rivers Avon and Stour, and the Moors River valley consist of extensive deposits of alluvium and the soils are calcareous and non-calcareous loams and clayey loams. Water in the River Avon and the Moors River is derived from chalk areas to the north, and when winter flooding occurs over the meadows an ample supply of calcareous matter is deposited, providing suitable habitats for calcicolous species. In the Moors River valley there are also some deep acid peat soils, and in these areas some very wet heath and bog habitats are present. Around Christchurch Harbour there are gley soils with fresh, brackish and salt marsh habitats, and along the coast from Bournemouth to Hengistbury Head and at Mudeford there are also several areas of blown sand deposits.

Figures 7 and 8 show the grid squares where heathland and woodland respectively can be found. Figure 8 also shows the grid squares with coniferous trees and plantations, and those with only deciduous woods and copses present. The amount of deciduous woodland cover in the area is extremely small, and although many grid squares indicate deciduous cover, the area of trees present is minimal and restricted mainly to small copses, shelterbelts and strips of remnant woodland.

Comparison of the geological and habitat maps show the relationship between the areas of heathland and the outcrops of the Branksome and Boscombe Sands. The coniferous plantations are almost entirely on the Branksome Sands, and the deciduous woods are generally on the Barton Clays and also in the main river valleys.

The large number of different habitats present in this small area of Dorset is the main contributory factor to the interest and diversity of the local flora. The juxtaposition of heathland, bogs and coniferous woodland with flood-meadows and neutral and acid grasslands, enhances the diversity of the flora. The riverine habitats and the arable land with hedgerows and roadside verges create links between many of the main habitats. In addition, the salt, brackish and fresh water marshes around Christchurch Harbour, and the sandy cliffs and dunes of the coast, provide further ecological variety. The dearth of large deciduous woodlands and lack of chalk downland in this area is more than compensated for by the abundance and richness of the other types of habitat.

CLIMATE

Bournemouth and the surrounding areas have an equable climate with a mean average annual temperature of 10.26^0C, however 1992 was the fifth successive warm year, with an above average mean annual temperature of 10.74^0C. The annual average number of sunshine hours is 1750. The average total annual rainfall is 778mm, although 1992 was the first year of above average rainfall since 1986, and also had the lowest annual total of sunshine hours since then. This was more noticeable as the previous three years had particularly fine summers (BEE, 9.1.93, p25). The lack of very cold winters with few occurrences of sleet and snow, and the generally mild temperatures and moderate rainfall provide suitable conditions for a wide variety of plant species to survive.

SCOPE AND METHOD OF SURVEY

The botanical records in this Flora are the result of fieldwork carried out between 1981 and 1993, and consist entirely of plants seen by the author in the field. Some plants were found by other people who showed them to the author, and others informed the author of locations and species which were subsequently searched for.

Appendix 2 lists additional native species recorded since 1980, by other botanists in the locality.

The arrangement and nomenclature of the species follow that of Clapham, Tutin and Moore, 3rd Edition, 1989, although some aggregate species follow the BRC plant names computer list. English names are generally according to Dony, Jury and Perring (1986), with a few as in Stace (1991) or CTM (1989). The majority of the area covered by the Flora is within vice-county 11, with the exception of seven grid squares in vice-county 9 (see Figure 2), and records were

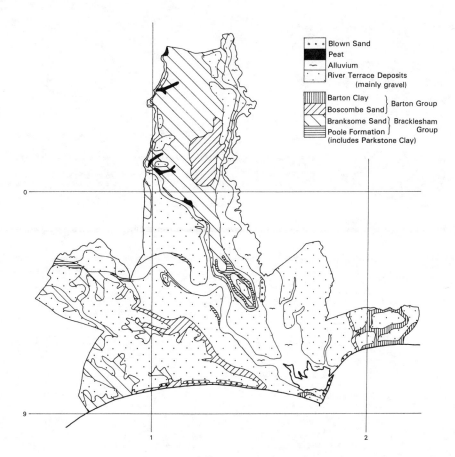

Blown Sand
Peat
Alluvium
River Terrace Deposits
(mainly gravel)
Barton Clay ⎱ Barton Group
Boscombe Sand ⎰
Branksome Sand ⎱ Bracklesham
Poole Formation ⎰ Group
(includes Parkstone Clay)

FIGURE 5. Surface geology of the area - simplified

FIGURE 6. Principal rivers and tributaries

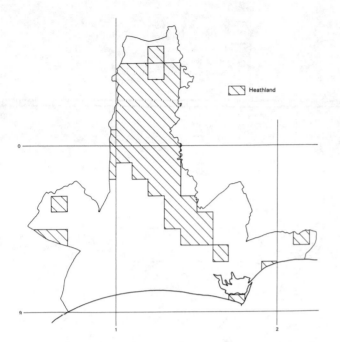

FIGURE 7. Grid squares containing heathland

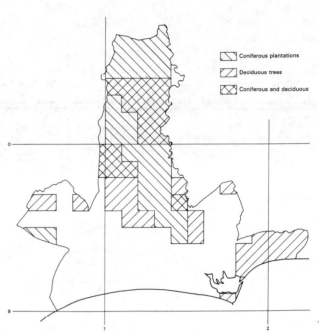

Coniferous plantations
Deciduous trees
Coniferous and deciduous

FIGURE 8. Grid squares containing deciduous and coniferous woodland

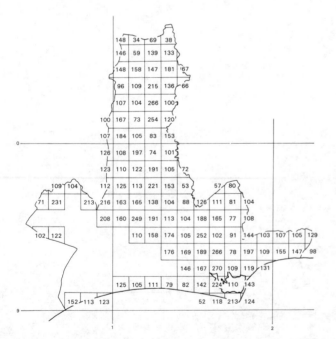

FIGURE 9. Number of species recorded in each grid square

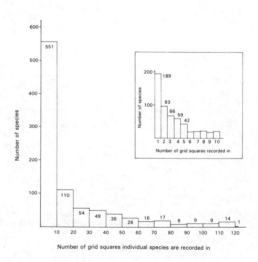

FIGURE 10. Number of grid squares in which individual species have been recorded

only included if they were within the boundaries shown, even though the grid square may continue beyond. The records collected are organised on a 1 kilometre National Grid square basis, and each grid square was visited on two occasions or more, and generally in at least two different seasons. The main habitats within each grid square were visited wherever possible, although some areas of private land have not been totally covered.

Gardens generally have not been included, although garden escapes, naturalized species, introductions and casuals are included. Forestry trees in general have not been included, except when an individual species was particularly noticeable or unusual. Individual *Rubus* and *Rosa* species are outside the scope of this book.

The inclusion of escapes, introductions and casuals provides a much more complete picture of the current flora of the area in the 1980s, and regardless of the means by which they have arrived, they are part of the vegetation cover and are growing alongside the native species of the area. Sites such as Bournemouth Cliffs consist of a mosaic of native, planted, introduced and casual species and escapes, and often it is impossible to discern the category to which some species belong.

It is also important to record the introduction and possible spreading of species such as *Agrostemma githago*, so that future records can be interpreted correctly, and where plants such as *Anacamptis pyramidalis* have been rescued and relocated, again it is vital to keep records for future studies. In cases where there was doubt about the correct identification, specimens were sent to recognised experts for confirmation. These included BSBI recorders and referees, and appropriate staff at The Natural History Museum in London. Although it is tempting to leave them out, some records have been included which have not been confirmed. Usually this is because either the specimen was not adequate, or the plant fell between two species descriptions or the plant could not be relocated for an 'authority' to see in situ. It may be of interest to others, and in the future, to know that there is a chance of these species being found, or refound, and hopefully they will be confirmed eventually.

SPECIES DISTRIBUTION MAPS AND DESCRIPTIONS

The botanical records are shown as distribution maps for each species. These maps were produced on an IBM-PC compatible computer using DMAP, a program supplied by Dr A Morton.*

Some maps show clearly a particular habitat distribution, for instance, *Alisma plantago-aquatica* and *Phragmites australis* follow the routes of the main rivers, and *Ammophila arenaria* and *Beta maritima* show a coastal distribution. The map of *Acer campestre* shows clearly that almost all records are from the Stour valley.

A few species show particularly interesting distributions, for instance *Cochlearia danica* which is found along the coast and also on the central reservation of the A338 road from Bournemouth to Ashley Heath. These are fully described in the individual species notes and distribution maps in the main part of the Flora.

* Dr. Alan Morton can be contacted at the Department of Biology, Imperial College, Silwood Park, Ascot, Berks SL5 7PY

When studying the distribution maps it must be remembered that a 'dot' in a grid square indicates only the presence of that species, and does not indicate the number of sites or quantity of plants for that species within each square. This is particularly important when studying the more uncommon species.

Maps and locations of the rarer and more vulnerable species have not been included as unfortunately there is always a risk to some species from unscrupulous collectors and other destructive influences. This can be illustrated no better than by the statement of Mansel-Pleydell concerning *Eleocharis parvula* at Little Sea, Studland: "abundant, and no chance of being extirpated by greedy collectors." (Mansel-Pleydell, 1895, p.287) It has not been seen there since 1936.

Maps are not included of species recorded in only one grid square, and some aliens and garden escapes are also not mapped. Maps shown of aggregates do not include the individual species which are mapped separately.

Species with almost complete coverage on the distribution maps might be assumed to be complete but this is not necessarily so. *Pteridium aquilinum*, for example, is recorded from 111 grid squares out of a possible maximum of 122, and could well be assumed to be present in almost all of them, but not one single piece has yet been recorded in grid square 1195, Throop Mill. This is surprising, and is certainly not due to lack of effort in searching for it.

The great majority of grid squares have over 100 species recorded in them, see Figure 9. Those with fewer are mainly either 'edge' squares or those along Bournemouth Cliffs where only a very small proportion of the square is vegetated and remains undeveloped. The other squares with under 100 species are not surprisingly the arable areas to the east of the River Avon and some of the sandy heathland areas. Higher numbers of species in squares following the major rivers are most likely due to the presence of several habitat types in the areas, and not related to the number of visits.

Inevitably some places have been visited considerably more than others, and it is difficult to determine whether the very high numbers of species recorded in a few squares is purely the result of a greater number of visits, or because of the extremely high botanical interest of those particular areas. The latter is the more likely, due mainly to the mosaic of habitats present.

The descriptions of individual species are arranged as follows:

The Latin name, authority and English name are given, followed by a local frequency rating showing the proportion of squares in which the species is recorded. The total number of grid squares is 122 and the categories are:

very common everywhere	recorded in over 100 grid squares
very common	80 - 99
common	60 - 79
fairly common	40 - 59
frequent	20 - 39
infrequent	4 - 19
uncommon	1 - 3
rare	regionally or nationally rare
very rare	very rare nationally

After this the general habitat is given, followed by notes on the local distribution. Additional notes, the previous distribution and whether the species is declining or increasing are then given where relevant. For species which have not been mapped the number of 1 kilometre grid squares in which they have been recorded is shown, followed by the 10 kilometre grid square reference. Species which are certainly introduced are preceded by an asterisk. Plants discovered by other botanists and subsequently intentionally visited by the author are described as 'seen'. Locations referred to by Linton and Townsend have been used with their spellings unaltered. Modern spellings have been used for locations of recent records.

DISCUSSION

Obviously any work covering an area of this size, undertaken by one person, can be nowhere near complete, and all the omissions and also any mistakes are solely the author's responsibility. It is hoped that the contents of the Flora will go some way towards refuting the statement by Good (1984) mentioned earlier (page 7, paragraph 4) that the additions to the Dorset flora of this part of Dorset would probably not be many. There are over 900 species, subspecies and hybrids present in just this small part of the county, several of them are only found here, and many are mainly found in this area and are uncommon in the rest of Dorset. However, the main concern should not be whether the plants are 'additional' to the rest of Dorset's flora, but whether we can retain the species which are still present here. The growth of Bournemouth and the loss of habitats, and changes in farming practice have resulted in the decline of many species already.

In this Flora it is shown that many species are found in only a few sites or areas, even allowing for a certain amount of under-recording, and it is sobering to realise that 30% of the species recorded in the area are present in only one or two grid squares. Figure 10 shows the numbers of grid squares in which species have been recorded, and one of the main conclusions that can be drawn is that a vast proportion of our species are now extremely vulnerable in this area. It is hoped that this Flora will act as a stimulus for other people in the area to go out and collect more records, especially where there are noticeable omissions. The additional information combined with other historical records could provide a considerable amount of invaluable information on which to base future study and decisions, which may help to safeguard the immensely varied flora present in this part of Dorset.

PTERIDOPHYTA

LYCOPODIACEAE

Lycopodiella inundata (L)Holub.
Marsh Clubmoss
Rare nationally; Dorset and the New Forest are the main UK centres. A single site with a few plants on a boggy heath with surface water. Has decreased considerably since 1900 when Linton described it as common on the heaths around Bournemouth and Christchurch. Less than 15 sites now in Dorset (Mahon & Pearman, 1993).
1 grid square;19

EQUISETACEAE

Equisetum fluviatile L.
Water Horsetail

Infrequent; ditches and wet meadows by rivers. Common in very wet meadows by the River Avon, and two records by the River Stour. Recorded as common by Linton and Townsend but generally much less frequent now.

Equisetum x litorale Kuhlew ex Rupr.
Shore Horsetail
Uncommon; the hybrid of *Equisetum arvense x E. fluviatile*. A single colony recorded in 1984, growing with both parents in a damp wooded chine.
1 grid square;29

Equisetum arvense L.
Field Horsetail

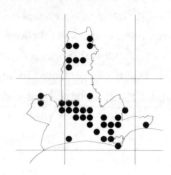

Frequent; hedgerows, fields and waste places. Noticeably more common in hedgerows and verges of the Stour valley and much less common on the drier sandy soils and heathlands.

Equisetum palustre L.
Marsh Horsetail

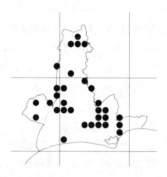

Frequent; marshes and wet boggy places. Common in wet meadows of the main river systems, and also by streams, tributaries and ditches. Recorded by Linton and Townsend in the Bourne Chine on the slipped clay cliffs.

Equisetum telmateia Ehrh.
Great Horsetail

Uncommon; damp woods, banks and marshy places. Two sites, in Chewton Bunny and Walkford, both in damp places near the edge of woodland. No other sites are mentioned in old Floras for this area.

OPHIOGLOSSACEAE

Ophioglossum vulgatum L.
Adder's-tongue

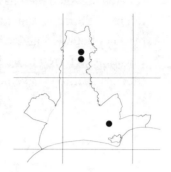

Uncommon; damp grassland and old meadows. Recorded at Purewell Meadows and in 1985 at two sites in Avon Forest Park. Two of the colonies have much decreased since then, and cannot be found every year. Recorded by Linton in meadows by the River Avon, but much more scarce now.

Ophioglossum azoricum C.Presl.
Small Adder's-tongue
Very rare nationally; short turf near the sea. A group of colonies in dry sandy grassland, inland. These have been identified by The Natural History Museum (London) as "probably" *O. azoricum*, but no chromosome count has been made. Only one other inland site is known, in the New Forest.
2 grid squares;10

OSMUNDACEAE

Osmunda regalis L.
Royal Fern
Frequent; wet peaty woods and heaths. Several sites on wet peaty soils at the sides of the floodplains of the Moors River and River Avon, and some scattered plants on the cliffs from Bournemouth to Highcliffe. Most records are at the edge of the Branksome and Boscombe Sands, or on peat. Almost all records for Dorset are in southeast Dorset and the Poole Basin.
21 grid squares;09,10,19

POLYPODIACEAE

Polypodium vulgare L. sens.lat.
Polypody

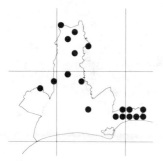

Infrequent; woods and hedgebanks. Widely scattered throughout the area, but much more common to the east of Christchurch. Previous Dorset records are more frequent in the west of the county.

Polypodium interjectum Shiras
Intermediate Polypody
Uncommon; woods and shady places. A single record from Avon Forest Park North in 1985. Undoubtedly under-recorded, some records are included with *P. vulgare*. Not distinguished by Linton or Townsend.
1 grid square;10

HYPOLEPIDACEAE

Pteridium aquilinum (L.)Kuhn.
Bracken

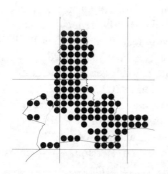

Very common everywhere; throughout the area, although not recorded in a few grid squares.

ASPLENIACEAE

Asplenium scolopendrium L.
Hart's-tongue

Infrequent; woods and shady hedgebanks. Several sites scattered throughout the area in woodland and copses. Much less frequent here than in the west of Dorset.

Asplenium adiantum-nigrum
L. sens.lat.
Black Spleenwort
Uncommon; on walls and hedgebanks. One single plant on a wall at the top of East Cliff, Bournemouth in 1987. By 1990 it had been "cleaned off" by council workers. Previously locally frequent, and so presumably declining.
1grid square;09

Asplenium ruta-muraria L.
Wall-rue

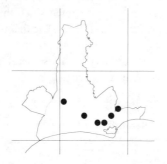

Infrequent; on old walls and bridges. Two sites on old railway bridges, and three sites on old walls. Also still found growing on the old Iford Bridge, a site mentioned in Linton and Townsend. Although supposedly common and widely distributed there are very few sites in the area, even though suitable walls are available.

ATHYRIACEAE

Athyrium filix-femina (L.)Roth
Lady-fern

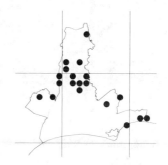

Infrequent; damp hedgebanks and woods. Several scattered sites in damp shady woods, but now not nearly as common around Bournemouth as described in Linton and Townsend.

ASPIDIACEAE

Polystichum setiferum
(Forsk.)Woynar
Soft Shield-fern

Infrequent; woods and hedgebanks. Several records in deciduous woods from Ensbury Wood to Chewton Bunny, but nowhere many plants. Also recorded in a forestry plantation at Ashley Heath. Described by Linton as frequent, and by Townsend as common, but apparently reduced in frequency now.

Asplenium ruta-muraria

Dryopteris filix-mas (L.)Schott
Male-fern

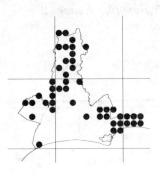

Fairly common; woods, hedgerows, scrub and wasteland. Widespread throughout the area, and generally throughout Dorset, but more commonly recorded in western and northeast Dorset, Purbeck and the Poole Basin.

Dryopteris affinis subsp. affinis
(Lowe)Fraser-Jenkins
Western Scaly Male-fern

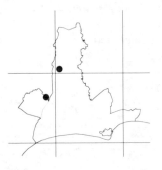

Uncommon; deciduous woods and hedgerows, also along rides in coniferous woodland. One site with several plants in Ensbury Wood, North Bournemouth, a damp deciduous woodland, also at St. Leonards. The _affinis_ agg. seems to be recorded more commonly in the west of the county.

Dryopteris affinis subsp. borreri
(Newman)Fraser-Jenkins
Borrer's Scaly Male-fern

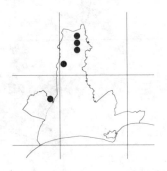

Infrequent; woods and hedges, usually on heavy soils and clays. Three sites for this subspecies in deciduous woodland and two in coniferous plantations. The Ensbury Wood site, a damp oak wood, has many plants with other species and subspecies. Previous records of the subspecies are unreliable, due to recent advances in the understanding of the taxonomy of this genus. Additional records will undoubtedly follow.

Dryopteris carthusiana
(Villar)H.P.Fuchs
Narrow Buckler-fern

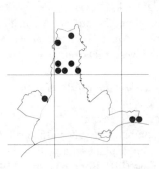

Infrequent; wet woodlands, marshes and wet heaths. Two records from wet heathland, several from wet deciduous woods and a few in coniferous woodland. Records in Dorset are from the Poole Basin and heaths and woods north of Bournemouth. Described by Townsend as rather common, it seems to have declined since then. Jermy & Camus (1991) state that it is "sensitive to seral succession and artificial drainage", it is therefore, vulnerable in particular on the wet heathland sites.

Dryopteris dilatata (Hoffm.)A.Gray
Broad Buckler-fern

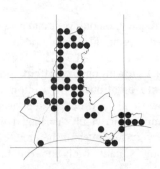

Fairly common; woods, hedgebanks and wet heaths. Widespread in woods throughout the area and generally found throughout Dorset, but more commonly recorded in west Dorset, the Poole Basin and eastern Dorset.

BLECHNACEAE

Blechnum spicant (L.)Roth
Hard Fern

Infrequent; damp woods and shady hedgebanks. All sites in shady woodland, on heathy or peaty ground. Also recorded from western Dorset and the Poole Basin. Possibly less common now than previously.

AZOLLACEAE

*Azolla filiculoides Lam.
Water Fern

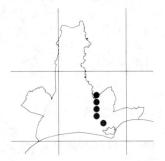

Introduced, infrequent; in rivers and ditches. Fairly frequent in the River Avon, and often abundant in ditches nearby. First mentioned by Rayner who listed it near Christchurch, in a pond in 1920 and in 1923 in ditches towards Sopley. Also recorded at Rempstone, Dorset in 1917 (Good, 1948), and in Good (1984) also at Ower and Herston. An increasing species, but so far only found regularly in the Avon valley.

GYMNOSPERMAE

PINACEAE

*Abies grandis (D.Don)Lindley
Giant Fir
Introduced,uncommon; planted for forestry and ornament. Two sites in Avon Forest Park South, both planted, one with many specimens mixed with other species in a small plantation on a heathy-grass field.
2 grid squares;10

*Pseudotsuga menziesii
(Mirbel)Franco
Douglas Fir
Introduced, uncommon; often planted for timber and ornament. Two sites, one in a Forestry Commission area near Hurn, the other in a shelter belt of trees on farmland near Winkton. There are many other unrecorded planted trees on forestry land throughout the area.
2 grid squares;19

*Picea abies (L.)Karsten
Norway Spruce
Introduced, uncommon; usually planted. A single record from a copse by the River Stour at Throop. Probably originally planted.
1 grid square;19

*Tsuga heterophylla (Rafin.)Sarg.
Western Hemlock-spruce

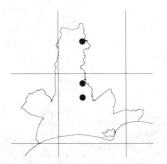

Introduced, uncommon; sometimes planted for timber. One tree in an old garden area in Avon Forest Park, and two recorded sites beside nature reserves in forestry plantations. Other Forestry Commission sites have not been recorded.

*Larix decidua Miller
European Larch

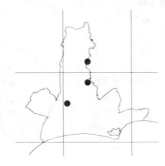

Introduced, uncommon; often planted. Three sites, two definitely planted and the third, below Matchams Viewpoint in Avon Forest Park probably also originally planted, but not recently. Forestry Commission plantations have not generally been recorded.

*Cedrus libani A.Rich.
Cedar of Lebanon
Introduced and planted, uncommon. One mature specimen planted in grass fields near the Bowling Green, Highcliffe. Trees planted in gardens have not been included.
1 grid square;29

Pinus sylvestris L.
Scots Pine

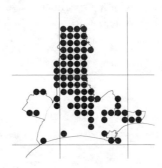

Very common; heaths and on light sandy soils. Widely planted in Bournemouth and Christchurch after the 'Christchurch Enclosure Act' of 1802, having already been tried experimentally by Sir George Tapps, Lord of the Manor of Westover at the end of the Eighteenth Century. Four 'Scots firs' at Heron (Hurn) Court may have been planted as early as 1746, thirty years before the species was reintroduced to the New Forest (Morris 1914). Also recorded by Mr Jennings at Barnfield, near Ringwood in 1787, in plantations then over 20 years old, giving a date

for introduction of before 1767 (Kenchington, 1944). Linton and Townsend recorded it (c. 1900) around Bournemouth, Herne and St Catherine's Hill. Since then a large amount of self-seeding has taken place.

*Pinus nigra Arnold
Corsican Pine

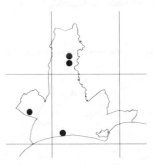

Introduced, infrequent outside plantations. Recorded from an old plantation in Avon Forest Park, and also a record from the cliffs at Boscombe in 1987. *Pinus nigra subsp.laricio* recorded in Avon Forest Park and *subsp. nigra* recorded at Turbary Common. Records from current forestry plantations have not been included.

*Pinus pinaster Aiton
Maritime Pine

Introduced, frequent; sandy heaths and near the coast. Introduced in Bournemouth in 1805, imported from the Landes, France, as it was more capable of withstanding drought than *Pinus sylvestris*. A sale notice of 1816 for land at Barnsfields (near Ringwood) mentions Pineasters planted about 15 years previously, giving an approximate date of introduction as 1801 (Driver, 1816).

*Pinus contorta
Douglas ex Loudon
Shore Pine (includes Lodgepole Pine)
Introduced, uncommon; sometimes planted. Two records in Avon Forest Park in 1985, one planted, the other perhaps planted although not known to have been.
2 grid squares;10

*Pinus radiata D.Don
Monterey Pine

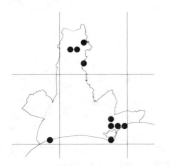

Introduced, infrequent; often planted. Several records in and around Avon Forest Park, and also many mature trees east of Christchurch.

*Pinus strobus L.
Weymouth Pine
Introduced, uncommon; occasionally planted. A single mature tree, originally planted, amongst other pines in Avon Forest Park North and recorded in 1985.
1 grid square;10

TAXODIACEAE

*Sequoiadendron giganteum
(Lindley)Buchholz
Wellingtonia
Introduced, uncommon; usually planted. A mature tree recorded in 1985 in Avon Forest Park, on the site of an old garden.
1 grid square;10

CUPRESSACEAE

*Chamaecyparis lawsoniana
(A.Murray)Parl.
Lawson's Cypress
Introduced, uncommon; planted for ornament and shelter. Two records, one in Avon Forest Park amongst many different large tree species, presumably originally planted as garden specimens for an old house.
2 grid squares;10

*Cupressus macrocarpa Hartweg
Monterey Cypress
Introduced, uncommon; often planted near the sea. Several mature trees at Hengistbury Head, planted in the old nursery area.
1 grid square;19

ARAUCARIACEAE

*Araucaria araucana
(Molina)C.Koch
Monkey Puzzle
Introduced, uncommon; planted. Two large trees planted in Avon Forest Park. The one in South park is not far from Matchams House. In North Park the tree is planted with many other single specimen trees of different species. Garden specimens have not been recorded.
2 grid squares;10

TAXACEAE

Taxus baccata L.
Yew

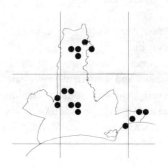

Infrequent, introduced locally, usually planted. Some trees obviously planted, for example in Holdenhurst Churchyard; and others present in copses and woodland, presumably at one time planted.

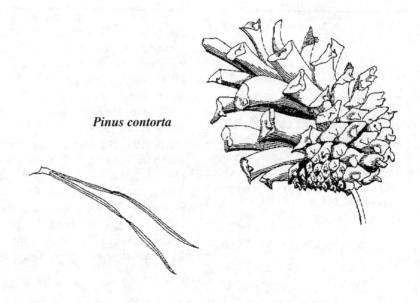

Pinus contorta

22

ANGIOSPERMAE - DICOTYLEDONES

RANUNCULACEAE

Caltha palustris L.
Kingcup, Marsh-marigold

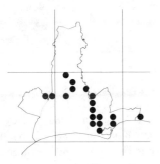

Infrequent; marshes and wet meadows. Common in wet places in the Avon, Stour and other river valleys, although not as widespread as might be expected.

*Nigella damascena L.
Love-in-a-Mist
Introduced, uncommon; garden escape. A single record from grassland by a woodland strip, Kinson.
1 grid square;09

Aconitum napellus L. sens.lat.
Monk's-hood
Uncommon; on shady stream banks in southwest England. A single plant growing on a steep bank on the edge of a woodland strip, near a stream in Kinson in 1990. Quite possibly a garden escape.
1 grid square;09

*Consolida ambigua (L.)Ball & Heywood
Larkspur
Introduced, uncommon; formerly in cornfields and as a casual or garden escape. One plant found flowering, and a smaller plant with leaves only, in 1987 near Throop, Bournemouth. The plants were growing with poppies on a bare patch of roadside verge where a trench had been dug for several weeks and then filled in. It is more than likely that the plants arose from seeds dormant in the ground, and not from gardens nearby. Plants were not found in 1988, and the verge was, by then, grassed over.
1 grid square;19

Anemone nemorosa L.
Wood Anemone
Uncommon; deciduous woods and hedgerows. One site only in deciduous woods in the east of the area. Surprisingly, no other sites recorded, although neither Townsend nor Linton list any sites in this area.
1 grid square;29

Clematis vitalba L.
Traveller's-joy, Old Man's Beard

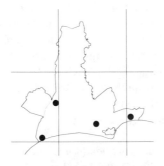

Infrequent; hedgerows and thickets, mainly on calcareous soils. One site at Middle Chine on the cliffs, some by the River Stour in hedgerows at Muscliff and at Wick. Also at Highcliffe Castle. Previously recorded at Iford Bridge by Linton.

Ranunculus acris L.
Meadow Buttercup

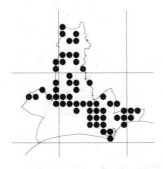

Fairly common; damp meadows and pastures. Recorded throughout the area, particularly in the Stour and Avon valleys.

Ranunculus repens L.
Creeping Buttercup

Very common everywhere; meadows, damp grassland, woods and waste ground. Probably present in every grid square.

Ranunculus bulbosus L.
Bulbous Buttercup

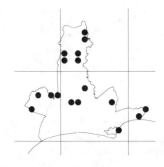

Infrequent; dry grassland and pastures. A few scattered records, but not nearly so common as indicated in old local Floras.

Ranunculus lingua L.
Greater Spearwort
Uncommon; marshes and fens. Recorded in 1991, planted in the new pond by the Visitor Centre at Avon Forest Park. Previously recorded however, by Linton and Townsend from several sites along the Moors River.
1 grid square;10

Ranunculus flammula L.
Lesser Spearwort

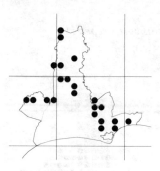

Frequent; wet and boggy places. Recorded in the Avon, Stour and Moors River valleys.

Ranunculus sceleratus L.
Celery-leaved Buttercup

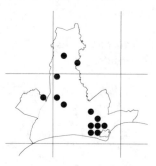

Infrequent; muddy ponds and ditches. Mainly in the lower Avon valley, but also by the River Stour. Particularly common on Stanpit Marsh.

Ranunculus hederaceus L.
Ivy-leaved Crowfoot

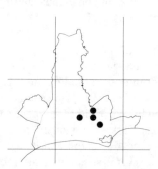

Infrequent; on mud and in shallow pools, streams and ditches. Three sites in meadows in the Avon valley and one near the River Stour. Recorded by Linton and Townsend as common in the Stour valley, possibly under-recorded now, but many sites must have disappeared since then.

Ranunculus omiophyllus Ten.
Round-leaved Crowfoot
Uncommon; wet muddy pools and ditches, usually in non-calcareous places. A single record from a wet muddy meadow by the Moors River near Hurn, in 1984. Previously recorded around Bournemouth, and on Sopley Common in 1879 (Townsend).
1 grid square;19

Ranunculus tripartitus DC.
Three-lobed Crowfoot
Rare; muddy ditches and pools where water stands temporarily. An unconfirmed record from the muddy edges of a farm pond in Parley (in VC9) in 1987. No recent records in Dorset, although recorded from Verwood in 1956 (Good, 1970), and on Stanpit Marsh in 1973 (Mahon & Pearman, 1993).
1 grid square;09

Ranunculus baudotii Godron.
Brackish Water-crowfoot
Rare; brackish water near coasts. Recorded in brackish pools on fresh to saltwater marshes. Few other sites locally, and not listed for this area in Linton or Townsend.
2 grid squares;19

Ranunculus aquatilis L. sens.lat.
Common Water-crowfoot

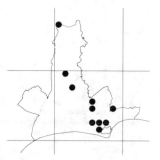

Infrequent; ponds, streams and ditches. Mainly in the lower Avon valley, and some sites in the Moors River valley.

Ranunculus peltatus Schrank.
Pond Water-crowfoot

Uncommon; ponds, ditches and shallow streams. Recorded on Priory Marsh, Stanpit in 1983, and by the River Mude in 1988. Previously recorded by Linton and Townsend by the Stour, and in ditches by the River Avon.

Ranunculus penicillatus
(Dumort.)Bab.
Stream Water-crowfoot

Infrequent; in moderate to fast-flowing rivers. Recorded in several sites in the Rivers Avon and Stour and the Moors River and also in the River Mude.

Ranunculus ficaria L.
Lesser Celandine

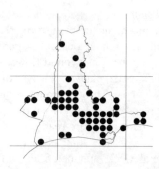

Fairly common; damp woods and shady hedgebanks. Common in the river valleys and flood plains, but not generally on the heathland soils.

Myosurus minimus L.
Mousetail
Rare; damp arable fields and grassland. Most sites are on road verges and tracks by arable fields, although tracks used by cattle seem to be good habitats. One grazed meadow in 1991 and 92 had an estimated 10-20,000 plants growing very lushly, quite remarkable for this generally declining species. Sporadic and rare in Dorset. Previously recorded near Christchurch, Ensbury and Kinson (Linton). Nationally declined from 59 to 16 10km grid squares (excluding casual records) between 1930 and 1990 (Wilson, 1992). 4 grid squares;19

Thalictrum flavum L.
Common Meadow-rue

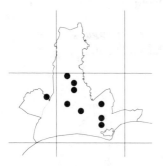

Infrequent; wet meadows, marshes and river banks. Recorded in the Avon and Stour valleys and the Moors River valley, but never very frequent, and usually not in large amounts. Recorded by Townsend as common along the Stour, and not frequent around Christchurch.

NYMPHAEACEAE

Nymphaea alba L.
White Water-lily

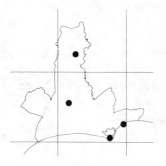

Infrequent; lakes, ponds and slow-moving streams. In the River Stour at Throop, and in ponds at Steamer Point,

Avon Forest Park and Hengistbury Head. The red waterlily on Hengistbury Head is also *N. alba* or a hybrid. Recorded by Linton and Townsend as common in the Stour and in the Avon, but less common now than then.

Nuphar lutea (L.)Sm.
Yellow Water-lily

Frequent; ponds, streams and slow-flowing rivers. Common in the River Stour and the Moors River, also recorded in the River Avon.

CERATOPHYLLACEAE

Ceratophyllum demersum L.
Rigid Hornwort

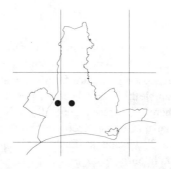

Uncommon; ponds and ditches. Both records from the River Stour, one of the locations very close to the Heron Court site mentioned in Linton and Rayner. Also previously recorded in ditches near Sopley and Christchurch in 1925 (Rayner). Possibly still more widespread than the current records suggest, as aquatics usually remain submerged and out of sight.

PAPAVERACEAE

Papaver rhoeas L.
Common Poppy

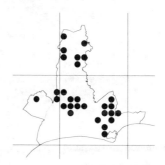

Frequent; a weed of arable fields and cultivated land, also in waste places and waysides. Mainly recorded around the arable fields east of the River Avon, in the cultivated meadows of the Stour valley and in the dry sandy grasslands in St Leonards parish.

Papaver dubium L.
Long-headed Poppy

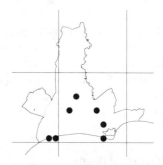

Infrequent; a weed of arable land and waste places. Several sites on the dry sandy cliffs between Bournemouth and Hengistbury Head.

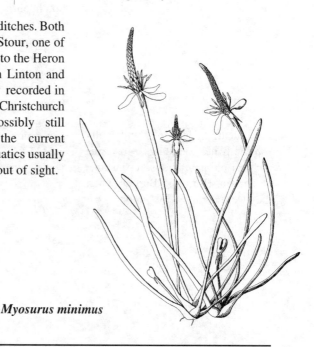

Myosurus minimus

***Papaver somniferum* L.**
Opium Poppy

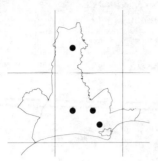

Introduced, infrequent; an escape on waste ground or a relic of cultivation. All four records are the subspecies *P. somniferum*, with pale lilac petals. Two sites on waste heaps, the other two on rough grassland and a roadside verge. Not recorded by Linton or Townsend in this area, although it was formerly often grown in many areas for medicinal purposes.

***Meconopsis cambrica* (L.)Vig.**
Welsh Poppy
Introduced, uncommon; casual, damp shady places. A few plants in Walkford in 1989, growing in a hedgerow with *Doronicum pardalianches* and *Geranium endressii*. Probably from dumped garden material, but may still be present.
1 grid square;29

***Chelidonium majus* L.**
Greater Celandine

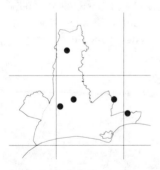

Infrequent; hedgebanks and walls, especially near buildings. Most sites are near old habitations, for instance at Throop, Hurn and Burton, where they were recorded by Linton.

***Eschscholzia californica* Cham.**
Californian Poppy

Introduced, uncommon; often a garden escape. One site on Bournemouth East Cliff, and two other sites where it probably arose from garden refuse.

FUMARIACEAE

***Corydalis claviculata* (L.)DC.**
Climbing Corydalis

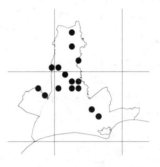

Infrequent; woods and shady hedgebanks. A few woodland sites, but most records are from scrub and bramble thickets on heathy commons, where it is present generally only in small quantities. Recorded by Linton and Townsend at Ensbury, Herne, Southbourne and Christchurch.

***Corydalis lutea* (L.)DC.**
Yellow Corydalis
Introduced, uncommon; on old walls and sometimes cultivated. Recorded from an old wall in Holdenhurst village, probably an escape from cultivation. Few records mentioned in old Floras for this area.
1 grid square;19

Fumaria muralis subsp. boraei
(Jord.)Pugsl.
Common Ramping-fumitory

Infrequent, possibly overlooked; cultivated and waste ground, hedgerows. All sites within three miles of each other, and all but one in the floodplain of the River Stour. Previously recorded from Iford Bridge to Wick, and Highcliff in 1894 (Linton).

***Fumaria officinalis* L.**
Common Fumitory

Infrequent; a weed of cultivated ground and waste places. Scattered records throughout the area, mainly on the lighter soils.

CRUCIFERAE

***Brassica oleracea* L.**
Wild Cabbage
Uncommon; coastal cliffs. A single record in 1993 of a large plant in dunes below the cliff at Hengistbury Head. Not previously recorded in this area, although frequent from Swanage to White Nothe and Portland.
1 grid square;19

*Brassica napus L.
Rape, Swede

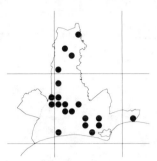

Introduced, frequent; waysides and cultivated ground. Often growing as an escape or relic of cultivation as this species 'oil seed rape' is grown for fodder, and for oil. Most common in the Stour valley.

Brassica nigra (L.)Koch.
Black Mustard

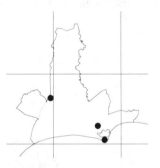

Uncommon locally; waste places and stream banks. Three records, one by the River Stour in 1989, one on a disturbed area of Stanpit Recreation Ground in 1990, and one at Hengistbury Head in 1993. Mentioned by Linton in the Stour valley and by Christchurch Harbour.

Sinapis arvensis L.
Charlock

Infrequent; cultivated and waste ground. Recorded from cultivated land, especially to the northeast of Christchurch, and along the coast.

*Hirschfeldia incana
(L.)Lagreze-Fossat
Hoary Mustard
Introduced, uncommon; grassy and sandy places. A single record from Southbourne in 1991 on a grassy roadside verge. Not mentioned in old Floras for Hampshire or Dorset. 1 grid square;19

*Diplotaxis muralis (L.)DC.
Annual Wall-rocket

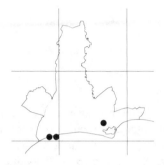

Introduced, uncommon; arable and waste ground, and on limestone rocks and walls. All three sites with few plants on sandy cliffs, but with limestone walls not far away in two cases. First recorded in Hants by Dr Trimen in 1863 as a weed in Bournemouth Upper Pleasure gardens (Townsend).

Raphanus raphanistrum L.
Wild Radish

Frequent; cultivated and waste ground. Generally recorded in cultivated ground in the Stour valley and northeast of Christchurch, but also common on Bournemouth cliffs.

Raphanus maritimus Sm.
Sea Radish

Infrequent; sandy cliffs and beaches. All four records from sandy cliffs, often growing with, and less common than, R. raphanistrum. Only recorded from Mudeford by Linton and Townsend, where it seems to have now disappeared. Few records from the Dorset coast.

Crambe maritima L.
Sea-kale
Rare; sand and shingle coasts. A single plant in 1991 growing on shingle at Hengistbury Head. Previously sporadically recorded between Christchurch and Mudeford by Townsend and Linton. Uncommon in Dorset, now mainly on the western coasts, and especially on Chesil Beach. 1 grid square;19

Cakile maritima Scop.
Sea Rocket

Rare; sand and shingle sea-shores. Scattered plants in several locations around Hengistbury Head, growing usually just above the drift-line. Certainly not still abundant as described by Linton. Few other Dorset sites now.

Lepidium campestre (L.)R.Br.
Field Pepperwort
Uncommon; dry places, roadsides, fields and waste ground. A single record in 1993 on a roadside verge at Ashley Heath. Previously recorded by Linton and Townsend at Highcliff and in the Avon valley.
1 grid square;10

Lepidium heterophyllum Bentham
Smith's Pepperwort

Uncommon; dry banks and fields. A few plants at each site, growing in dry gravelly soil. Certainly not now as frequent or common as described in Linton and Townsend.

Coronopus squamatus
(Forsk.)Aschers.
Swine-cress

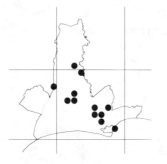

Infrequent; waste ground and trampled places. Most records are from meadows in the Stour and Avon valleys, usually where the gateways are trampled by cattle.

*Coronopus didymus (L.)Sm.
Lesser Swine-cress

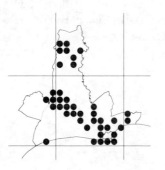

Introduced, fairly common; cultivated and waste ground. Mainly in the Stour and Avon valleys, more frequent than *C. squamatus*, although described in most Floras as less common.

*Cardaria draba (L.)Desv.
Hoary Cress

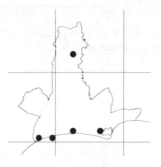

Introduced, infrequent; cultivated land and disturbed ground. Five sites for this weed species, one at Avon Forest Park, the others on Bournemouth East and West Cliffs, Boscombe cliffs and Hengistbury Head. A record for Bournemouth cliffs is mentioned in Linton. Described in various Floras as a troublesome weed and spreading well, although there is no evidence of this in the local area at present.

Thlaspi arvense L.
Field Penny-cress
Uncommon; waste places and arable land. A single record from fields near Muscliff in 1982, not far from the River Stour. Recorded as common by Linton and Townsend around Iford and downstream by the River Stour, also from the Avon valley. Infrequent in Dorset (Good, 1984).
1 grid square;09

*Thlaspi alliaceum L.
Garlic Penny-cress
Introduced, very rare; a weed of arable land. A single record from a track by arable fields near Winkton in 1987. Not previously recorded in local Floras; less than 10 'dots' nationally since 1950 (Rich, 1991), and none of these are in Dorset or Hants.
1 grid square;19

Teesdalia nudicaulis (L.)R.Br.
Shepherd's Cress
Uncommon; on sandy banks. A single site on a dry sandy roadside bank, with several plants present. Linton and Townsend recorded several sites in the area; recorded in Good (1984)

at only one other site in Dorset, at Sandbanks.
1 grid square;19

Capsella bursa-pastoris (L.)Medic.
Shepherd's-purse

Common; cultivated and waste ground. Widespread throughout the area, but more common on the river valley soils.

Cochlearia anglica L.
English Scurvygrass

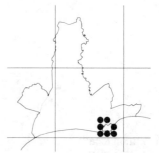

Infrequent; estuaries and muddy shores. Present all round Christchurch Harbour, growing in large quantities, and recorded there by Linton and Townsend.

Cochlearia danica L.
Danish Scurvygrass

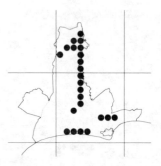

Frequent; sandy banks, dry places and walls near the sea. Recorded from a small stretch of cliffs from Boscombe Pier to Southbourne where it is quite plentiful. Also on the central reservation of the A338 between the Cooper Dean roundabout and Ashley Heath and the A31 to Tricketts Cross, and on the Christchurch by-pass. It has presumably spread from the coast, associated with the higher concentration of salt following road-gritting in winter. Almost always only recorded from the central reservations, and not on the sides of the same roads. Not listed by Linton or Townsend for this area.

Cochlearia officinalis L. sens.str.
Common Scurvygrass
Uncommon; saltmarshes, shingle, rocks and cliffs by the coast. Several plants recorded in 1990 growing on shingle and blown sand on the cliffs at Hengistbury Head. Not previously recorded for this area in old Floras. Although common in the rest of the British Isles, it is rare on southern and southeastern coasts.
1 grid square;19

Lunaria annua L.
Honesty

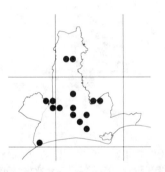

Introduced, infrequent; a garden escape. Scattered records throughout the area in a variety of habitats, but most common in the Stour valley.

Lobularia maritima (L.)Desv.
Sweet Alison

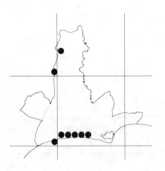

Introduced, infrequent; a garden escape, sometimes naturalized.
Many plants present on Bournemouth cliffs, mostly naturalized, and two records inland.

Erophila verna (L.)Chevall sens.lat.
Common Whitlowgrass

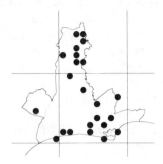

Frequent; dry banks, grassy places and tracks. Scattered throughout the area, especially on dry sandy and gravelly tracks.

Armoracia rusticana
Gaertn.Mey.& Scherb.
Horse-radish

Introduced, infrequent; an escape or relic of cultivation. Several sites, often on roadside verges near old established villages such as Throop, Purewell and Winkton.

Cardamine pratensis L. sens.lat.
Cuckooflower

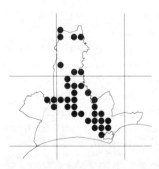

Frequent; damp meadows and wet places. Recorded from meadows in the Stour and Avon valleys, and in the Moors River valley, but not recorded from several other smaller stream and riversides. Not nearly as common now as previously, mainly due to loss of suitable meadows and use of fertilizers.

Cardamine flexuosa With.
Wavy Bitter-cress

Frequent; moist shady stream banks. Most records are from the banks of the River Stour and the Moors River, with scattered records elsewhere. Much less common than *C. hirsuta*.

Cardamine hirsuta L.
Hairy Bitter-cress

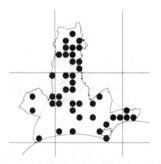

Fairly common; in dry places, on hedgebanks and walls, and as a weed. Widely distributed throughout the area.

Barbarea vulgaris R.Br.
Winter-cress

Frequent; damp hedges, stream and river banks. Common on the banks of the River Stour, and some records from the Moors River and River Avon.

Barbarea verna (Mill.)Aschers.
American Winter-cress
Introduced, uncommon; waste ground and waysides. A single record in 1992 from Avon Forest Park South, in

grassland on the site of an old building. Recorded by Linton and Townsend about Bournemouth and at Christchurch.
1 grid square;10

Arabis hirsuta (L.)Scop.
Hairy Rock-cress
Very uncommon in this locality; dry banks and walls.
A few plants growing on the dry calcareous area of a new overflow car park on Stanpit Recreation Ground, in 1988. Recorded at Southbourne in 1920, no other records in old Floras for this area.
1 grid square;19

Rorippa sylvestris (L.)Besser
Creeping Yellow-cress

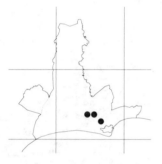

Uncommon; riversides and wet meadows. Two records in meadows by the River Avon and below Christchurch, and one record from wet meadows by the River Stour near Iford. Only recorded by Rayner near Christchurch, no other local sites previously mentioned in old Floras.

Rorippa palustris (L.)Besser
Marsh Yellow-cress

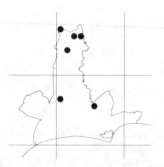

Infrequent; wet places and ditch banks. Recorded from banks of the Rivers Stour and Avon, and from the Moors River and one of its tributaries. Probably under-recorded from these river banks, but still not common in this area.

Rorippa amphibia (L.)Besser
Great Yellow-cress
Rare; very few sites known regionally. A single site with a few plants only, by the River Stour, at Throop in 1981. Also recorded by Linton and Townsend at Throop, the only site mentioned.
1 grid square;19

Nasturtium officinale R.Br. agg.
Water-cress

Frequent; ditches, streams and wet places. Common in the Avon valley, also recorded from the Stour, Mude and Moors River valleys.

Nasturtium microphyllum
(Boenn.)Reichenb.
Narrow-fruited Water-cress
Uncommon; streambanks and ditches. A single record from the banks of the Moors River in 1988. Undoubtedly under-recorded, and possibly as common as *N. officinale*. Only a few previous records in Dorset and Hants, recorded at Throop by Linton and Townsend.
1 grid square;19

Matthiola incana (L.)R.Br.
Hoary Stock
Very rare as a native, but often introduced by the sea. A single plant recorded in 1992 on a shingle beach at Mudeford, probably introduced, but not apparently cultivated. Not listed by Linton or Townsend for this area.
1 grid square;19

*Hesperis matronalis L.
Dame's-violet
Introduced, uncommon; a garden escape in meadows, hedgerows and waysides. A single record from grassland by the River Stour at Muscliff, Bournemouth in 1989. Several records in Rayner but not mentioned in Linton or Mansel-Pleydell.
1 grid square;09

Alliaria petiolata (Biebe)Cavara & Grande
Garlic Mustard, Jack-by-the-Hedge

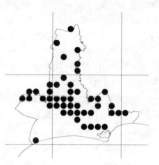

Fairly common; hedgerows and shady places.
Most common in the hedgerows of the river floodplains, especially the Stour, but recorded throughout the area.

Sisymbrium officinale (L.)Scop.
Hedge Mustard

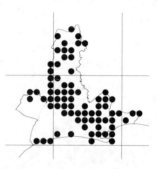

Common; hedgebanks, roadside verges and waste places. Widespread throughout the area, but more common on the heavier soils.

Arabidopsis thaliana (L.)Heynh.
Thale Cress

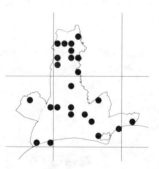

Frequent; dry banks and waste places. Several records from dry banks in the Stour and Avon valleys, and also the grassy heathlands near St. Leonards.

RESEDACEAE

Reseda luteola L.
Weld

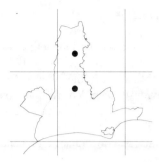

Uncommon; roadsides and waste places, often but not always on calcareous substrata. One record in 1985 by a concrete road near Avon Forest Park, and one in 1993 on Sopley Common. Not recorded by Linton in this area, but recorded by Townsend near Mudeford.

VIOLACEAE

Viola odorata L.
Sweet Violet

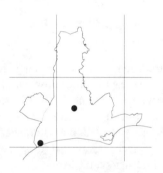

Uncommon; woods and hedgebanks. Recorded in 1983 from a small copse in Holdenhurst village, and in Middle Chine in 1992. Although generally common in Dorset and Hants there are few previous records for this area.

Viola riviniana Reichenb.
Common Dog-violet

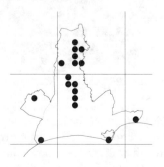

Infrequent; woods, hedgebanks and heaths. Recorded from woods and shady places mainly on the heathland soils north of Hurn.

Viola canina L.
Heath Dog-violet

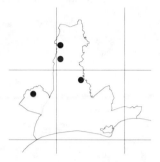

Uncommon; heaths and dry grassland. All four sites on heathland and heathy grassland. Recorded by Linton and Townsend from many sites around Bournemouth, most of these sites have since disappeared.

Viola lactea Sm.
Pale Dog-violet

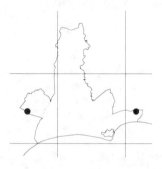

Uncommon; heathland. Recorded in 1989 on Chewton Common, and in 1993 on Turbary Common. Few records in southwest Hampshire (Townsend) and about eight sites in Dorset (Mahon & Pearman, 1993). Recorded by Linton and Townsend on heaths and commons around Bournemouth, and at Kinson and Chewton. Much less frequent now.

Viola tricolor L.
Wild Pansy

Rare; cultivated and waste ground, grassland. A single plant in 1986 on disturbed ground in an overgrazed horse field, Broadway Lane, Throop, and one on the old Iford Bridge in 1992. Recorded by Linton in fields near Wick and Throop, not listed elsewhere in this area.

Viola arvensis Murray
Field Pansy

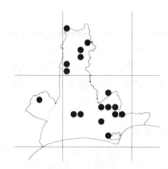

Infrequent; cultivated and waste ground. Most frequent on the arable land to the east of the River Avon, but also recorded on disturbed sandy grassland.

POLYGALAEAE

Polygala vulgaris L.
Common Milkwort

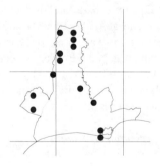

Infrequent; in grasslands, but more usually on calcareous grassland. Recorded on the sandy grasslands around St Leonards, also on Kinson Common and Hengistbury Head. Recorded by Linton in the Avon meadows near Christchurch.

Polygala serpyllifolia Hose
Heath Milkwort

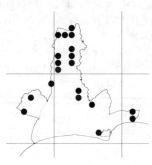

Infrequent; sandy heaths and grassy places on acid soils. Recorded from most of the sandy grasslands in the area, more common than *P. vulgaris.*

HYPERICACEAE

Hypericum androsaemum L.
Tutsan

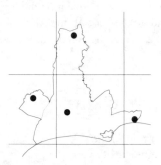

Infrequent; woods and hedgerows. Recorded at Ashley Heath beside the disused railway line, on Kinson Common, in North Bournemouth and at Chewton Bunny. Chewton is the only site listed in Linton and Townsend for this area.

Hypericum perforatum L.
Perforate St John's-wort

Frequent; grassland, hedges and woods. Recorded throughout the area, but mainly on the dry sandy grasslands and heaths north of Hurn. Not recorded by Townsend in this area

but described by Linton as common off the heaths. Especially common on calcareous soils according to CTM and Good, but recorded in this survey mainly on sandy grassland and heaths.

Hypericum tetrapterum Fries
Square-stalked St John's-wort

Infrequent; damp meadows and stream banks. Recorded from banks of the Moors River, the River Stour and the Avon valley. Probably present by most streams in the area and possibly much under-recorded.

Hypericum humifusum L.
Trailing St John's-wort

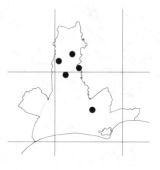

Infrequent; dry sandy heaths. A few small plants at each site, all on heathland with short sparse vegetation. Apparently much more common previously than now.

Hypericum pulchrum L.
Slender St John's-wort

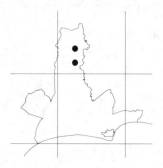

Uncommon; dry grass heaths. Two single plants found at separate sites in Avon Forest Park in 1985 and 1991.

Described as common and frequent by Linton and Townsend, but definitely more uncommon now than previously.

Hypericum elodes L.
Marsh St John's-wort
Uncommon; bogs and wet places on acid soils. Two sites only, both on heathland near the Moors River. Previously recorded as common in the Stour and Avon valleys by Linton and Townsend. A declining species, mainly due to drainage of suitable habitats.
2 grid squares; 10,19

TAMARICACEAE

*Tamarix gallica L.
Tamarisk

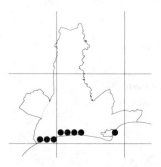

Introduced, infrequent; often planted near the sea. Recorded from the cliffs between Alum Chine and Southbourne, some probably originally planted, but naturalized in many places. Linton lists it as planted on Bournemouth Cliffs.

CARYOPHYLLACEAE

Silene dioica (L.)Clairv.
Red Campion

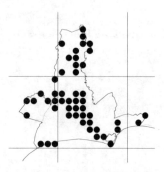

Fairly common; hedgerows, woods and shady places. Recorded throughout the area, but most common in hedgerows in the Stour valley.

Silene latifolia Poiret.
White Campion

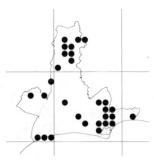

Frequent; hedgerows, grassy banks and waste places. Recorded from the Stour valley, but more frequent in hedgerows around arable land north of Christchurch, and on grassland near St Leonards.

Silene vulgaris subsp.vulgaris
(Moench)Garke
Bladder Campion

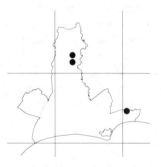

Uncommon; hedgerows and dry grassland. A few plants, in scattered locations, but no large colonies. Recorded as rather local and not common in this area by Linton and Townsend, but more records in Dorset.

Silene gallica L. sens.lat.
Small-flowered Catchfly
Rare; sandy and gravelly fields and waste places, often a casual. A single site on sandy cliffs, with several patches of well established plants, seen in 1989. This species which may be native, but is often introduced, was previously frequent and is now decreasing nationally. Recorded by Linton and Townsend at Kinson, Heron Court, Wick, Southbourne and Highcliff in 1888, also at Christchurch in 1923 (Rayner).
1 grid square;19

Lychnis flos-cuculi L.
Ragged-Robin

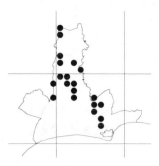

Frequent; damp meadows and marshy places. Common in the Avon valley and the Moors River valley, but less so in the Stour valley. Previously common in the Stour valley (Linton) and probably decreasing due to loss of habitat, drainage and increased use of herbicides and fertilizers.

Lychnis flos-cuculi

*Lychnis coronaria (L.)Desr.
Rose Campion

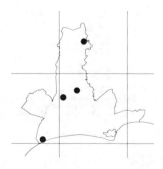

Introduced, infrequent; a garden escape. Several plants growing on the edge of a car park at Hurn, also in rough ground in Durley Chine, and at Throop and Ashley Heath. All sites probably from dumped garden material.

*Agrostemma githago L.
Corncockle
Planted; an introduced weed of cornfields. Four plants grown from seed in a wildflower packet in a corner of the playing field at Epiphany School, Muscliff in 1989. A further plant at the front of the school in 1991 was grown from seed from the 1989 flowers. Extremely rare as an arable weed. Previously recorded in the Stour watershed and at Highcliff, Burton and Christchurch (Linton, Townsend).
1 grid square;09

Dianthus armeria L.
Deptford Pink
Very rare nationally; dry sandy fields and pastures. One site found in 1986, with a few plants growing in a dry sandy field. In 1987 several more plants were found about 100 yards away, in long grass and patches of heather, and three further plants nearby in 1992. Not listed for this area in old Floras. This is the only extant site in VC 11, and the only recent record in Dorset. Nationally declined from 43 10km squares since 1930 (ABF) to about 27 in 1960 (BRC Monkswood, pers.comm.) and still decreasing.
1 grid square;10

Dianthus deltoides L.
Maiden Pink
Rare; dry grassland, but often a casual or garden escape. One site with a good colony of plants, and a few additional plants scattered nearby, recorded in 1985. These plants may have been escapes, or planted many years ago. The other site consists of a few small plants growing wild in a garden, about 1 mile from the main site, and also present before 1985. Not mentioned in old Floras for this area.
1 grid square;10

Saponaria officinalis L.
Soapwort
Uncommon; hedges and waysides near villages, usually an escape. A colony recorded in a grassy clearing by a road to scattered dwellings near Woolsbridge. Previously recorded near Bournemouth and Christchurch by Linton and Townsend.
1 grid square;10

Cerastium fontanum Baumg.
Common Mouse-ear

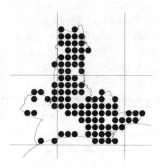

Very common everywhere; meadows, hedgerows and grassy places throughout the area.

Cerastium glomeratum Thuill.
Sticky Mouse-ear

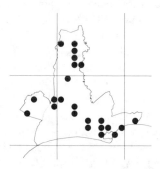

Frequent; fields, banks and dry places. Recorded from grassland in the Stour and Avon valleys, and also in dry sandy grassland areas elsewhere.

Cerastium diffusum Pers.
Sea Mouse-ear

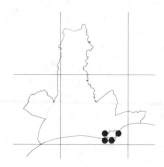

Infrequent; sandy and stony places, especially near the coast. All four records on dry sandy grassland near the sea. Probably under-recorded as other sites would be expected. Recorded by Linton as commonest near the sea, but also recorded at Knapp Mill and Hurn (Linton, 1919). In Dorset it has spread inland on main road verges.

Cerastium semidecandrum L.
Little Mouse-ear

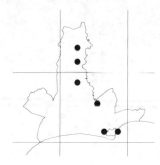

Infrequent; dry sandy grassland. Recorded from several sites, all on short dry sandy grassland. Probably under-recorded as there are many other suitable sites, such as the cliffs at Bournemouth. Previously recorded there, and at Redhill, about Christchurch, and Hengistbury (Linton, Townsend).

Myosoton aquaticum (L.)Moench.
Water Chickweed

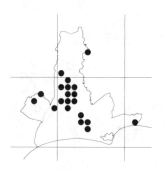

Frequent; wet places and streamsides. Common in the Moors River valley, and also at places in the Stour valley, but recorded from few other places in the area.

Stellaria media (L.)Vill. agg.
Common Chickweed

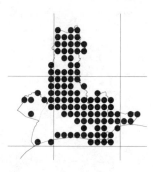

Very common everywhere; disturbed ground, roadsides and in grassland throughout the area.

Stellaria pallida (Dumort)Pire
Lesser Chickweed

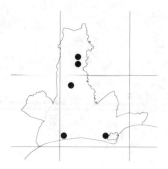

Rare; dry sandy grassland. Four sites recorded in 1991 and one in 1992. In almost all of these sites this species was growing with or amongst *S. media*, so it may well be rather overlooked elsewhere. A few other Dorset records recently.

Stellaria neglecta Weihe
Greater Chickweed
Uncommon; hedgerows, wood margins and shady places. A single site, under oak trees in a hedgerow on Purewell Meadows in 1982. Not now present as the area is much overgrown with bushes and brambles, due to lack of grazing. In Dorset, only common in the extreme west.
1 grid square;19

Stellaria holostea L.
Greater Stitchwort

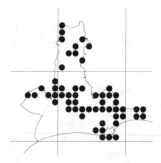

Fairly common; woods, hedgebanks and shady places. Recorded in most areas, but less common on the dry heathy soils.

Stellaria palustris Retz.
Marsh Stitchwort
Very rare; marshes and base-rich fens. Two sites only, with not very large populations, one site has few more than five plants. Previously found in the Stour and Moors River valleys and other meadows, therefore

obviously declining; although one of the present sites was first found in 1849 and still survives. 2 grid squares;19

Stellaria graminea L.
Lesser Stitchwort

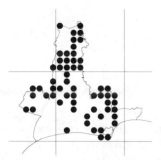

Fairly common; heaths, grassland and hedgerows. Recorded throughout the area, but more frequent on the drier sandy soils.

Stellaria alsine Grimm
Bog Stitchwort

Infrequent; streamsides and wet boggy meadows. Common in the Avon valley, but less so in the Stour valley. Previously common in the Stour valley (Linton), the decrease is probably due to drainage and loss of habitats.

Moenchia erecta (L.)Gaertn.Mey.& Scherb.
Upright Chickweed

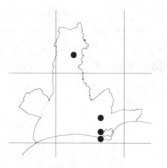

Rare; dry sandy grassland. Several good sites for this rapidly declining species. The site at Purewell Meadows

found in 1983 was destroyed by new housing in 1989. Previously described as not uncommon in the area by Linton, but declined considerably now and only recorded from three areas in the rest of Dorset since 1980.

Sagina apetala Ard. **subsp. apetala** (**S.ciliata** Fries)
Ciliate Pearlwort

Uncommon; sandy places and dry grassland. One record from Hengistbury Head in 1988, and one at Avon Forest Park in 1991. Probably under-recorded as it was recorded by Linton from several locations around Christchurch Harbour and elsewhere in the area.

Sagina apetala subsp. erecta (Hornem.)F.Hermann
Annual Pearlwort

Frequent; heaths and dry grassland. Recorded mainly around Christchurch Harbour, and on the sandy grasslands around St Leonards.

Sagina maritima G.Don fil.
Sea Pearlwort

Uncommon; shores and sandy grassland by the coast. One site on sandy cliffs by the sea, and one by a sandy shore on Stanpit Marsh in 1987. Previously recorded at Hengistbury, Mudeford, Christchurch and Bournemouth (Linton, Townsend).

Sagina procumbens L.
Procumbent Pearlwort

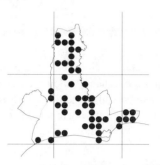

Fairly common; damp bare ground, tracks, grassland and cultivated land. Recorded throughout the area, but more frequent in the Avon and Stour valleys.

Sagina subulata (Swarz)C.Presl.
Heath Pearlwort
Rare; dry sandy heaths and grassland near the coast. A single record from dry sandy heathland on St Catherine's Hill in 1982. Few recent records in Dorset, probably overlooked but also decreasing nationally. Described by Linton and Townsend as generally distributed in the Stour watershed, at Holdenhurst, Throop and at Highcliff in 1888.
1 grid square;19

Sagina nodosa (L.)Fenzl.
Knotted Pearlwort
Uncommon; damp sandy and peaty places. A single record from Coward's Marsh, Christchurch in 1983. Recorded by Linton and Townsend from several heaths, and around Christchurch Harbour. Apparently decreased since then.
1 grid square;19

Honkenya peploides (L.)Ehrh.
Sea Sandwort

Uncommon; sandy and shingly seashores. All three sites around Christchurch Harbour, on sand and shingle beaches. Although described as common all round the British Isles (CTM) there are not many other sites on the Dorset and west Hants coast.

Moehringia trinervia (L.)Clairv.
Three-nerved Sandwort

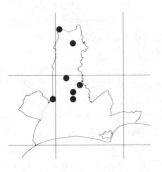

Infrequent; damp shady hedgebanks and woods. A few scattered records, mostly in woodland around Hurn.

Arenaria serpyllifolia L.
Thyme-leaved Sandwort

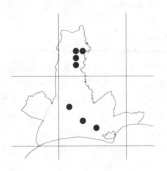

Infrequent; cultivated ground, waste places and dry sandy ground. Many records from sandy grassland in Avon Forest Park, and a few from dry grassy places in the Stour valley. Probably under-recorded.

Arenaria serpyllifolia subsp. leptoclados (Reichenb.)
Lesser Thyme-leaved Sandwort

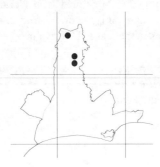

Uncommon; dry sandy grassland and bare ground. Three records, all on short dry heathy grassland areas. Probably much under-recorded and possibly more common than it appears.

Spergula arvensis L.
Corn Spurrey

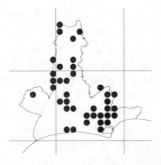

Frequent; arable fields and disturbed places. Most common in the arable fields to the east of the River Avon, but also recorded in dry fields in the Stour valley and around Parley.

Spergularia rubra (L.)J.& C.Presl.
Sand Spurrey

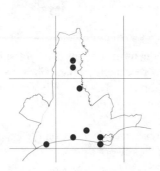

Infrequent; sandy and gravelly places. Recorded mainly from the sandy cliffs and cliff tops by the sea, but also at Avon Common and in Avon Forest Park on dry sandy grassland.

Spergularia rupicola Lebel ex Le Jolis
Rock Sea-spurrey

Uncommon; sandy cliffs and rocks by the sea. All three sites on dry sandy cliffs or cliff tops. Recorded by Linton and Townsend on cliffs at Bournemouth.

Spergularia marginata (DC.)Kittel
Greater Sea-spurrey

Infrequent; saltmarshes by the coast. Several records from Stanpit Marsh and other sites around Christchurch Harbour.

Spergularia marina (L.)Griseb.
Lesser Sea-spurrey

Infrequent; salt marshes. Recorded all around Christchurch Harbour.

PORTULACACEAE

Montia fontana L.
Blinks

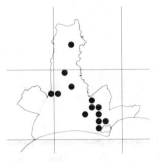

Infrequent; wet meadows and streamsides. Found frequently in the Avon valley, also recorded in the Stour and Moors River valleys.

*Montia perfoliata (Willd.)Howell
Springbeauty

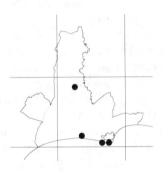

Introduced, infrequent; cultivated and disturbed ground especially on sandy soils. Several patches on Hengistbury Head. First noticed in Hants in 1856 in Bure Lane, Mudeford, abundant in 1879 and destroyed by housing before 1885 (Townsend). Also recorded at East Cliff in 1892 (Linton) and in Alum Chine (Linton, 1919). Only a few records in Dorset, in the Poole Basin.

AIZOACEAE

*Carprobotus edulis (L.)N.E. Br.
Hottentot-fig

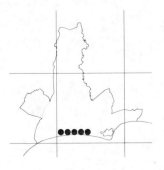

Introduced, infrequent; sandy cliffs by the sea. Growing in patches of various sizes on the cliffs between Bournemouth Pier and Southbourne. Large areas died during the cold winter of 1987/88, but most have since regrown. Not mentioned in any old local Floras, but reported from Southbourne Rd. in 1929 by A.W. Graveson (MS in Dorchester Museum Herbarium).

CHENOPODIACEAE

Chenopodium album L. agg.
Fat-hen

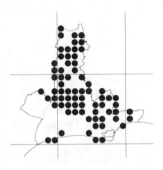

Common; cultivated ground and waste places. Widespread throughout the area.

Chenopodium ficifolium Sm.
Fig-leaved Goosefoot
Uncommon; waste and arable land, particularly round manure heaps. One site, on edge of overgrown land in Winkton. Recorded by Linton and Townsend as a weed in the public gardens, above Bournemouth Square, 1894-6.
1 grid square;19

Chenopodium rubrum L.
Red Goosefoot
Uncommon; waste ground, especially rubbish tips and farmyards. One record from East Parley, on a track leading to a disused farmyard. Good and CTM describe this plant as frequent, although Linton and Townsend both considered it to be rare.
1 grid square;19

Beta vulgaris subsp. maritima (L.)Thell.
Sea Beet

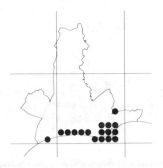

Infrequent; seashores and cliffs. Common around Christchurch Harbour and recorded from many places on Bournemouth cliffs.

Atriplex littoralis L.
Grass-leaved Orache

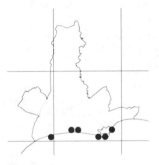

Infrequent; seashores and sandy cliffs. Recorded from six places on the coast, including around Hengistbury Head and at Mudeford, where it was recorded as frequent by Linton.

Atriplex patula L.
Common Orache

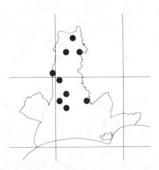

Infrequent; cultivated and disturbed ground. Several scattered records, in waste places and disturbed ground away from the sea. Probably under-recorded, described as very common by Linton and Townsend.

Atriplex prostrata Boucher ex DC.
Hastate Orache

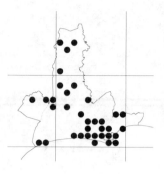

Frequent; cultivated ground and sandy ground near the sea. Most records from around Christchurch Harbour and along the coast, also recorded from the Stour valley and other scattered sites. 'Noticed' by Townsend in the Stour watershed and at Mudeford, Highcliff and Chewton.

Atriplex laciniata L.
Frosted Orache

Uncommon; sandy and shingly shores around the high-tide mark. Recorded in 1988 from Hengistbury Head, and in 1993 also on the south side. Recorded by Linton and Townsend in the same location as abundant in 1893, and also at Highcliff. In Dorset very occasional around Poole Harbour.

__Atriplex halimus__ L.
Shrubby Orache

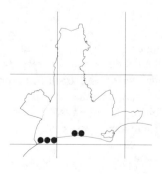

Introduced, infrequent; a garden escape and sometimes naturalized. Recorded from the cliffs between Alum Chine and Southbourne, probably planted and becoming naturalized. Excluded by Townsend and listed only on the Isle of Wight at Freshwater.

Halimione portulacoides (L.)Aellen
Sea-purslane

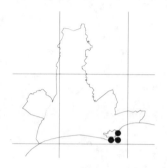

Uncommon; saltmarshes, especially in channels and pools flooded at high tide. Recorded in the saltmarshes below Hengistbury Head. Recorded by Townsend also at Mudeford. Generally found in saltmarshes along the coasts of Dorset and Hants.

Suaeda maritima (L.)Dumort
Annual Sea-blite
Uncommon; saltmarshes. Recorded in the saltmarshes at Hengistbury Head, but not common there. Previously also recorded at Christchurch and Mudeford (Linton, Townsend).
1 grid square;19

Salsola kali subsp. kali L.
Prickly Saltwort
Uncommon; sandy seashores. A single record on the sandy beach at Hengistbury Head in 1991. Previously also recorded by Linton and Townsend between Mudeford and Christchurch, and by Linton in plenty between Mudeford and Highcliff in 1901 (Townsend). No recent records in Dorset (Mahon & Pearman, 1993).
1 grid square;19

Salicornia agg.
Glasswort

Infrequent; saltmarshes. Recorded from several sites around Christchurch Harbour, including Stanpit Marsh and below Hengistbury Head. Not recorded from many places in Dorset as saltmarshes are few and far between, although usually common where it is found.

Salicornia ramosissima Woods
Purple Glasswort
Uncommon; upper parts of saltmarshes and muddy shores. Recorded from Stanpit Marsh South in 1987. Previous records of the distribution of species in this genus are confused and unclear, although this species was recorded on Stanpit Marsh (Linton, 1919) and from below Hengistbury Head in 1923 (Rayner). 1 grid square;19

Salicornia pusilla Woods
One-flowered Glasswort
Uncommon; drier parts of saltmarshes. One record, probably of this species, from the saltmarshes below Hengistbury Head in 1986. Previously recorded on Stanpit Marsh in 1903 (Townsend).
1 grid square;19

TILIACEAE

__Tilia platyphyllos__ Scop.
Large-leaved Lime

Introduced, infrequent; usually planted, probably native only in and around the Wye valley. Records of mature trees from three sites in the Stour valley and also near Highcliffe and on Turbary Common, presumably all originally planted.

***Tilia x europaea* L.**
Lime

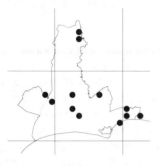

Introduced, infrequent; usually planted. Several trees in the Stour valley and near Highcliffe, but very few elsewhere.

MALVACEAE

***Malva moschata* L.**
Musk Mallow

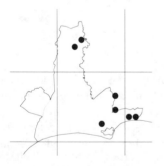

Infrequent; edges of woods, fields and hedgerows. Four scattered records in hedgerows, and three from scrubland and bushy places. Recorded by Linton as not unfrequent, but somewhat decreased since then.

***Malva sylvestris* L.**
Common Mallow

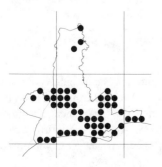

Fairly common; roadsides and hedgerows. Common in the Stour valley, along the coast and in hedgerows to the east of the River Avon, but very infrequent on the heathland soils.

***Malva neglecta* Wallr.**
Dwarf Mallow

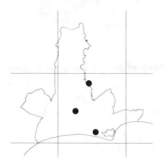

Uncommon; waste places and roadsides. One record from Wick Meadows in 1986, one from Holdenhurst in 1992 and in the Avon valley in 1993. Recorded often by Linton and Townsend in the Stour valley and lower Avon valley, but has obviously declined since.

***Lavatera arborea* L.**
Tree-mallow

Rare as a native, elsewhere introduced; sea coasts and cliffs. All sites are on seafront cliffs or shores, plants grow 'wild' but were probably originally introduced. These sites are not mentioned in Linton or Townsend.

***Althaea officinalis* L.**
Marsh-mallow

Rare; saltmarshes, mainly on south and east coasts. Several clumps on Stanpit Marsh. Stanpit and the Fleet area are the only two sites in Dorset, and also on the south coast west of Lymington. In 1900 Linton recorded it as very rare and possibly extinct but Townsend (1904) stated "Stanpit 3 or 4 strong clumps (Linton)". Linton listed it on Stanpit Marsh in 1922 (Linton, 1925).

***Linum bienne* Miller**
Pale Flax
Uncommon; dry grassland near the sea. A single plant in 1991 growing in a flowerbed on the cliff top between West Cliff and Durley Chine, Bournemouth. Probably not planted here, but may have been an escape from cultivation elsewhere. Recorded by Linton and Townsend from several sites in this area.
1 grid square;09

***Linum catharticum* L.**
Fairy Flax

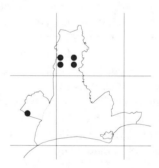

Uncommon; dry grassland, usually calcareous but also in sandy places. All sites amongst grass on sandy heathland. Linton described it as absent from heathland districts, and Townsend did not find it common in this area.

***Radiola linoides* Roth**
Allseed
Rare; damp sandy places on heaths. Two plants found in 1991 on Hengistbury Head on a grassy bank, fifty metres away from the site recorded previously by others. Few other sites in Dorset, a fast decreasing species. Described as rather common by Townsend, and Good noted 102 sites present in Dorset in 1948.
1 grid square;19

GERANIACEAE

Geranium pratense L.
Meadow Crane's-bill
Uncommon; roadsides and meadows. One plant growing on a roadside verge and field edge in 1981, near East Parley Church. Possibly a garden escape, but not particularly close to any gardens. Recorded by Townsend near Christchurch.
1 grid square;19

*Geranium endressii Gay
French Crane's-bill
Introduced, uncommon; casual of roadsides and hedgerows. A single plant in Walkford in 1989 growing in a hedgerow with *Doronicum pardalianches* and *Meconopsis cambrica*. Probably from dumped garden material. Not recorded in old Floras for this area.
1 grid square;29

*Geranium phaeum L.
Dusky Crane's-bill
Introduced, uncommon; a garden escape, sometimes naturalized. A single plant in 1988 growing on a grassy slope beside the disused railway line near Ashley Heath, probably a garden escape.
1 grid square;10

Geranium pyrenaicum Burm. fil.
Hedgerow Crane's-bill

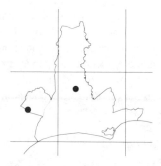

Uncommon; hedgerows and roadsides. One record on a roadside verge near Hurn Village in 1990, and one in 1993 on Turbary Common. Not recorded in this area by Linton or Townsend, but some sites in Dorset.

Geranium dissectum L.
Cut-leaved Crane's-bill

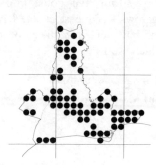

Common; hedgerows, grassland and waste ground. Most common in the Stour and Avon valleys and to the east of Christchurch.

Geranium molle L.
Dove's-foot Crane's-bill

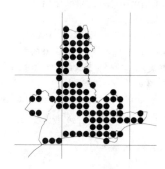

Very common; hedgerows, grassland, cultivated and disturbed ground. Widespread throughout the area.

Geranium pusillum L.
Small-flowered Crane's-bill

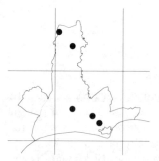

Infrequent; dry grassland and banks and verges with open habitats. Three sites in the Stour valley, two elsewhere. Recorded as frequent by Linton, and common locally by Townsend; although it is now much more common in eastern England than in the central and western areas.

Geranium lucidum L.
Shining Crane's-bill

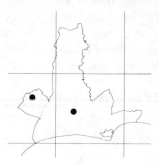

Uncommon; dry walls and hedgebanks. Two sites, one in Kinson and one on a dry wall and bank near Holdenhurst. Both plants are near buildings and it is not possible to say if either have been introduced or not. Not originally recorded by Linton or Townsend for this area, but found later in Moordown (Linton, 1919).

Geranium robertianum L.
Herb-Robert

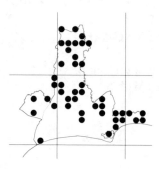

Fairly common; woods, hedges and shady places. Recorded throughout the area, but not as common as might be expected. Also a colony with white flowers recorded at Ashley Heath in 1993.

Erodium moschatum (L.)L'Herit.
Musk Stork's-bill

Uncommon; sandy, grassy and waste places. A record from the grassy bank of the River Avon, near Knapp Mill in 1988. Also an unconfirmed record in 1982 from a Water Company spoil heap a few hundred yards away. Very rare in Hants (Townsend) and uncommon in Dorset (Good, 1984). Only two post-1970 records in Dorset (Mahon & Pearman, 1993). Recorded in 1879 by the roadside at Chewton village (Linton, Townsend).

Erodium cicutarium (L.)L'Herit
Common Stork's-bill

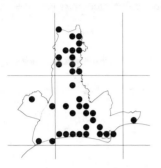

Frequent; dry sandy grassland. Recorded mainly on sandy cliffs along the coast and on the light sandy soils of the heathlands, few records elsewhere.

Erodium cicutarium subsp. bipinnatum (Willd.)Tourlet
Sticky Stork's-bill
Uncommon; sandy places near the sea. A record in 1993 on sandy ground at Hengistbury Head. A few records previously from Dorset, not mentioned by Linton or Townsend. 1 grid square;19

OXALIDACEAE

Oxalis acetosella L.
Wood-sorrel

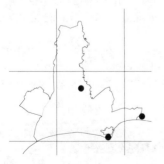

Uncommon; damp woods and shady places. Two large sites in damp deciduous woods, and a small colony on the edge of a plantation, but near woodland. Only recorded in a few sites in this area by Linton and Townsend, but more frequent in the rest of Dorset.

*Oxalis corniculata L.
Procumbent Yellow-sorrel
Introduced, uncommon; waste places and cultivated ground. One record in 1986 from waste ground off Woodbury Avenue, Bournemouth, probably from dumped garden material, and one near Mudeford. 2 grid squares;19

*Oxalis articulata Savigny
Pink-sorrel

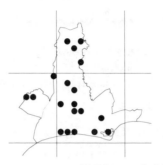

Introduced, infrequent; a garden escape. Scattered records throughout the area as an escape on waste and disturbed ground and on cliffs at Bournemouth.

*Oxalis corymbosa DC.
Large-flowered Pink-sorrel
Introduced, uncommon; a casual, sometimes naturalized. A single record in 1981 from a gateway on a roadside verge near Hurn, probably from dumped garden material. 1 grid square;19

BALSAMINACEAE

*Impatiens capensis Meerburgh
Orange Balsam

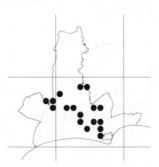

Introduced, infrequent; banks of rivers and streams. Common on the banks of the River Stour, and also by the Avon. First recorded in Britain in 1822 in Surrey. First mentioned in local Floras by Rayner at Wick Ferry and on the banks of the Avon in 1921. Recorded by Good (1948) along the

Stour below Bryanston. Probably spreading, but not quickly.

*Impatiens glandulifera Royle
Indian Balsam

Introduced, infrequent; riverbanks, streams and ditches. Several records from banks of the Stour, but also recorded on the sides of small streams and ditches. Two records in 1993 on Hengistbury Head. Not previously recorded in this area in local Floras. First mentioned in Good (1948) as uncommon, and in Good (1984) as frequent in Dorset. Could increase fairly quickly locally in the future.

TROPAEOLACEAE

*Tropaeolum majus L.
Nasturtium
Introduced, uncommon; a garden escape. Recorded in 1985 in Avon Forest Park North near the workbase, almost certainly a garden escape although not cultivated nearby. 1 grid square;10

ACERACEAE

*Acer pseudoplatanus L.
Sycamore

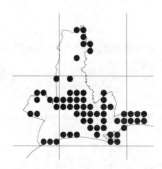

Introduced, common; woods and hedgerows. More common on the wetter soils, especially in the Stour valley and around Highcliffe. Recorded on Bournemouth cliffs, but less frequent on lighter sandy heathland soils.

* Acer platanoides L.
Norway Maple
Introduced, uncommon; often planted. A record on Turbary Common, Bournemouth; most probably planted. Other sites undoubtedly present, although not previously recorded in this area.
1 grid square;09

Acer campestre L.
Field Maple

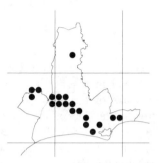

Infrequent; hedgerows and woods. Common in the Stour valley, but very infrequent elsewhere. Linton described it as "common (off the heathland)" but it now appears to be much more restricted.

HIPPOCASTANACEAE

*Aesculus hippocastanum L.
Horse-chestnut

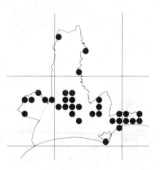

Introduced, frequent; usually planted. Most records from the Stour valley and around Highcliffe, but also planted in other places.

ANACARDIACEAE

*Rhus typhina L.
Sumach
Introduced, uncommon; a garden escape. One record from Winkton on an overgrown undeveloped plot, amongst trees.
1 grid square;19

AQUIFOLIACEAE

Ilex aquifolium L.
Holly

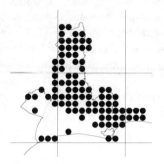

Very common; woods, hedgerows and scrub. Widespread throughout the area, but not in exposed places on the coast.

CELASTRACEAE

Euonymus europaeus L.
Spindle

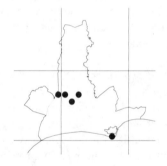

Infrequent; hedgerows and thickets. Several records from hedgerows in the Stour valley, and one record from the wood at Hengistbury Head. Described by Linton and Townsend as common, but much decreased now.

BUXACEAE

Buxus sempervirens L.
Box

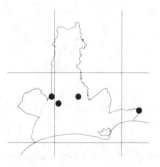

Infrequent; sometimes planted and becoming naturalized. Mostly in hedgerows not all that far from gardens. One site in woodland, which once belonged to the estate of a large house, so undoubtedly planted then. Only native on chalk from Glos. to Surrey, all local sites therefore planted at some time.

RHAMNACEAE

Rhamnus catharticus L.
Buckthorn

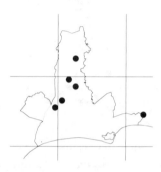

Infrequent; hedgerows and woods. A few scattered records only. Described as rare in this area by Townsend, and certainly still very infrequent.

Frangula alnus Miller
Alder Buckthorn

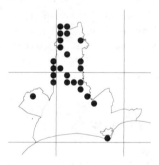

Frequent; damp heaths and peaty woods. Most records from damp places and ditches around heaths and commons, and woodland on the edge of peaty areas. Recorded by Linton and Townsend as common in the Stour watershed, but much decreased since then.

VITIDACEAE

*Parthenocissus quinquefolia
(L.)Planchon
Virginia Creeper
Introduced, uncommon; often planted, sometimes escapes. Recorded on the cliff tops at West Cliff, Bournemouth, straggling amongst other vegetation.
1 grid square;09

LEGUMINOSAE

Laburnum anagyroides Medicus
Laburnum

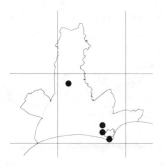

Introduced, infrequent; a garden escape. Records include a tree growing in a meadow by the Moors River, nowhere near any habitation, and a record from the old rubbish tip edge at Priory Marsh, Christchurch.

Cytisus scoparius (L.)Link
Broom

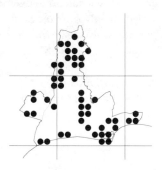

Fairly common; heaths and dry sandy soils. Recorded from most of the heaths in the area, and in several places on the cliffs.

Genista anglica L.
Petty Whin
Uncommon; on heaths. A few plants recorded in 1992 on a heathland site adjoining meadows by the River Avon. Only recorded by Linton in three sites in this area.
1 grid square;19

Ulex europaeus L.
Gorse

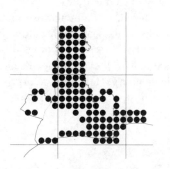

Very common everywhere; heaths and grassy places, usually on light, acid soils. Not recorded from the area of arable fields east of Burton.

Ulex gallii Planchon
Western Gorse

Uncommon; on heathland mainly in the West and strongly calcifuge. Three records, all on heathlands near Bournemouth, where it was recorded as abundant by Linton and Townsend. Possibly under-recorded, but not as frequent in this area as *Ulex minor*. In Dorset only frequent west of Weymouth and Dorchester.

Ulex minor Roth
Dwarf Gorse

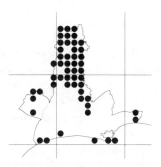

Fairly common; heathlands. Recorded from almost all heaths in the area, and on the sandy cliffs along the coast. Not recorded from the west of Dorset as the Poole Basin is on the western limit of its British range.

Lupinus polyphyllus Lindley
Garden Lupin
Introduced; a garden escape, uncommon. Several plants growing on dry shingle on Spellars Point, Stanpit Marsh in 1991.
1 grid square;19

Lupinus arboreus Sims
Tree Lupin

Introduced, infrequent; sand dunes, shingle and sandy cliffs. Common on sandy cliffs all along the coast, also on dunes at Hengistbury Head. Recorded in a few places inland, all fairly near the River Stour. The first record in Dorset was at South Haven in 1953 (Good, 1955).

Robinia pseudoacacia L.
False-acacia
Introduced, uncommon; usually planted. One record in 1987 in Hurn Forest, near St Leonard's Hospital, almost certainly planted; and another planted at Muscliff, Bournemouth. 2 grid squares;09,10

Vicia hirsuta (L.)S.F. Gray
Hairy Tare

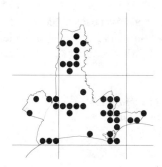

Fairly common; hedgerows and grassy places. Recorded from the Stour valley and on sandy grassland near the sea, but more common on the lighter soils east of the River Avon.

Vicia tetrasperma (L.) Schreber
Smooth Tare

Uncommon; grassy places and verges. All three records from damp grassland in 1987 and 1988. Recorded only in two locations in Linton and Townsend, although there are more records from the west of Dorset.

Vicia cracca L.
Tufted Vetch

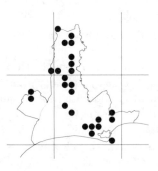

Frequent; hedges, grassland, bushy places and river banks. Recorded from the Stour and Moors River valleys, and from hedgerows and streambanks elsewhere.

Vicia sepium L.
Bush Vetch

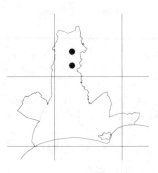

Uncommon; hedgerows and grassy places. Two records in 1984 on dry sandy grassland. Although common throughout the British Isles, apparently not frequent in this area, either now or previously. Townsend described it as 'apparently rare in the SW' and Linton as 'not common'.

Vicia sativa L. sens.lat.
Common Vetch

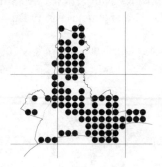

Very common; hedges, road verges and grassy places, often introduced. Widespread throughout the area, but not so common in wooded areas and plantations.

Vicia sativa subsp. nigra (L.)Ehrh.
Narrow-leaved Vetch

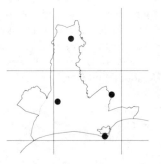

Infrequent; dry banks and grassy places. Four records from different habitats, one from verges by arable fields, two from dry sandy grassland and one on a disturbed grassy area. Previously common in the Stour and Avon watersheds and on dry grassy commons (Linton,Townsend). Much less common now, although some records may be included in *V. sativa* sens.lat.

Vicia lathyroides L.
Spring Vetch
Very rare; dry sandy turf. Two sites near Christchurch Harbour, one recorded in 1983; the other in 1991, with only a few plants, therefore very vulnerable. Only three sites recorded in total in the area in Townsend, Linton and Rayner, one of which was destroyed by buildings before 1885. 2 grid squares;19

Lathyrus nissolia L.
Grass Vetchling
Uncommon; grassy places. A single site with a few plants, on a grassy bank by the River Stour in 1989. By 1991 many plants were scattered over an area of several metres. Not recorded by Linton or Townsend for this area.
1 grid square;09

Lathyrus pratensis L.
Meadow Vetchling

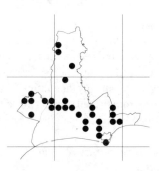

Frequent; grassland and hedgerows. Recorded mainly from the Stour and Avon valleys, also in the Moors River valley and by the River Mude.

*Lathyrus latifolius L.
Broad-leaved Everlasting-pea

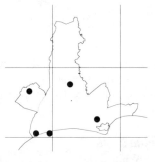

Introduced, infrequent; an escape, somtimes naturalized. A few scattered records, three probably originating from outcast garden refuse, and two records from the cliffs at Bournemouth where it appears to be naturalized.

Pisum sativum L.
Garden Pea, Field Pea
Introduced, uncommon; waste places, usually as an escape from cultivation. A plant with a single purple flower recorded in 1993, growing in sand on Hengistbury Head. A surprising record in such a hostile habitat.
1 grid square;19

Ononis repens L.
Common Restharrow

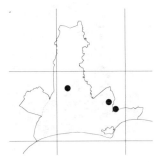

Uncommon; dry grassy places, especially on calcareous soils. Two sites on dry roadside verges north of Christchurch, one by a track at Merritown. Recorded by Linton as common in the Stour watershed but certainly not so now.

Melilotus altissima Thuill.
Tall Melilot

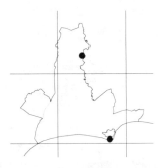

Possibly introduced; uncommon, more common on the chalk. One record on a roadside verge in Matchams Lane in 1986 and one at Hengistbury Head in 1993. No locations listed in old Floras, described as rare in this area.

*Melilotus officinalis (L.)Pallas
Ribbed Melilot

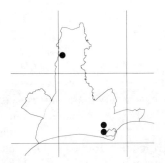

Introduced, uncommon; fields and waste places. All three sites with only a few plants, growing on waste ground and roadside verges. No records for this area in old Floras, and very few in Dorset.

*Melilotus alba Medicus
White Melilot
Introduced, uncommon; fields and waste places. A single record from a roadside verge near St Leonard's Hospital. Not recorded in this area by Linton or Townsend, and only a few records in Dorset.
1 grid square;10

Medicago sativa subsp. falcata
(L.)Arcangeli
Sickle Medick
Introduced (native only in East Anglia), uncommon; grassland and waste places. A single record in 1986 from the edge of the old rubbish tip by Priory Marsh, Christchurch. Recorded by Linton and Townsend on the railway bank north of Christchurch in abundance and by Rayner at Knapp Mill. Also recorded at Middle Chine (Good, 1955). A few Dorset records.
1 grid square;19

Medicago lupulina L.
Black Medick

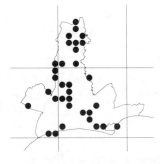

Frequent; roadsides, field borders and grassland. Recorded mainly from the Stour and Moors River valleys, but also along the coast and from the sandy grasslands in the north of the area. Generally more common in Dorset on the chalk areas to the west and north.

Medicago arabica (L.)Hudson
Spotted Medick

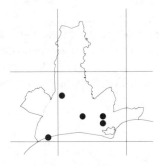

Infrequent; grassland, especially on sandy and gravelly soils. Recorded from five sites, three of them in the Stour valley, all in rough grassland. Townsend and Linton recorded it as frequent near the coast, and from several places between Bournemouth and Christchurch, and at Mudeford.

Trifolium ornithopodioides L.
Bird's-foot Clover, Fenugreek

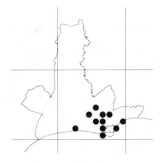

Infrequent; sandy and gravelly banks and grassland near the coast. All sites are on dry sandy grassland, some grid squares have several sites for this species which is not uncommon locally in the right habitat, although most colonies are not extensive. Two sites on Purewell Meadows were destroyed in 1989 by the 'new' housing estate. Some sites in the rest of Dorset on the coast and on sand inland.

Trifolium repens L.
White Clover

Very common everywhere; grassland and roadsides, especially on clayey soils. Almost certainly present in every grid square, the few gaps are most likely under-recorded.

*Trifolium hybridum subsp. hybridum L.
Alsike Clover

Introduced, infrequent; roadsides and field borders. A few scattered sites on roadside verges. Recorded by Townsend in Bournemouth in 1889, and described by Linton as on the increase, but not currently common in this area.

Trifolium glomeratum L.
Clustered Clover

Rare nationally; dry sandy or gravelly grassland, mainly near the sea. The two sites in Purewell Meadows in 1983 with reasonable colonies were destroyed in 1989 for a housing estate. Four sites remain, one with only a few plants; another, a colony growing as weeds on a sparse lawn in a large garden in Mudeford, found in 1991, and some plants recorded at Wick and on Hengistbury Head in 1993. A decreasing species, previously locally abundant (Linton), and now only nine other recent records in Dorset.

Trifolium suffocatum L.
Suffocated Clover

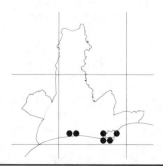

Rare nationally; dry, short, sandy grassland near the sea, where the grass cover is not too dense. Two sites found in 1987, one in 1991 and two in 1993, all near Christchurch Harbour. Also recorded from the cliff tops at Boscombe in 1993. Two of the sites were recorded originally before 1883, so have obviously survived for over 100 years. These two sites have large populations of this species, although the plants themselves are very small. Also recorded by Linton near Hurn. Only five other grid squares with post-war records in Dorset.

Trifolium fragiferum L.
Strawberry Clover

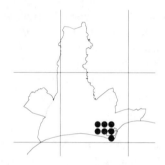

Infrequent; damp grassland near the sea. Recorded in saline grassland in many sites around Christchurch Harbour, but not recorded elsewhere in the study area. Previously recorded by Linton around Christchurch Harbour, along the coast and in the Avon valley. Also recorded at Throop (Linton, 1925).

Trifolium campestre Schreber
Hop Trefoil

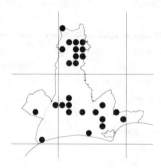

Frequent; dry grassland and roadsides. Recorded mainly in the Stour and Avon valleys and on the dry grasslands in the north of the area. Described as common by Townsend and Linton.

Trifolium dubium Sibth.
Lesser Trefoil

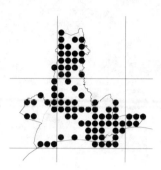

Very common; dry grassland, roadsides and disturbed ground. Recorded from most habitats, especially where there are patches of well-drained rough grassland.

Trifolium micranthum Viv.
Slender Trefoil

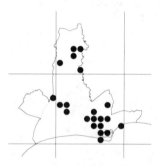

Frequent; dry sandy grassland. Mainly recorded on short sandy turf in the Avon and Stour valleys and around Christchurch Harbour.

Trifolium striatum L.
Knotted Clover

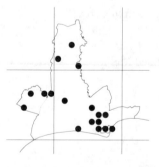

Infrequent; dry grassland, especially on sandy soils. Recorded from the Stour and Avon valleys and around Christchurch Harbour. One inland record from Avon Forest Park on dry sandy grassland.

Trifolium arvense L.
Hare's-foot Clover

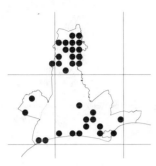

Frequent; dry grassland and sandy places. Present in many places along the coast, and a few records from the Stour valley. Recorded mainly on the dry grassland and sandy heaths of Avon Forest Park, and around St Leonards and Ashley Heath. Described by Townsend as very common in the Stour watershed, but much less common there now.

Trifolium pratense L.
Red Clover

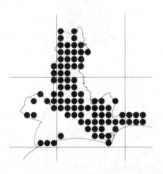

Very common; meadows, roadsides and grassy places. Widespread throughout the area, probably present in all grid squares.

Trifolium subterraneum L.
Subterranean Clover

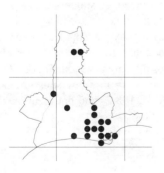

Infrequent; dry sandy grassland. Mainly recorded around Christchurch Harbour and in the Stour and Avon valleys, also several sites on sandy grassland near the sea. Recorded by

Linton as frequent in and around Bournemouth and Christchurch. Much more frequent here than in the west of Dorset.

Lotus corniculatus L.
Common Bird's-foot-trefoil

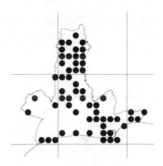

Fairly common; meadows, roadsides and grassy places. Recorded from most habitats throughout the area, although not found in the intensively farmed arable areas to the east of the River Avon. Not recorded as commonly as would be expected; previous Floras imply it is very common everywhere.

Lotus tenuis Waldst. & Kit. ex Willd.
Narrow-leaved Bird's-foot-trefoil
Uncommon; dry grassy places. A single record from Avon Forest Park in 1985, in an area of dry heathland with sandy bridleways. Very few previous records in this locality, Dorset records are mainly from around Poole Harbour and Weymouth. 1 grid square;10

Lotus uliginosus Schkuhr
Greater Bird's-foot-trefoil

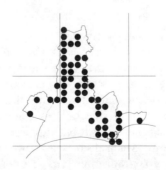

Fairly common; damp grassy places. Recorded throughout the area in damp grassy habitats, but with noticeable gaps in distribution.

Lotus subbiflorus Lag.
Hairy Bird's-foot-trefoil

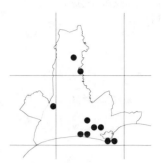

Infrequent; sandy and dry grassy places near the sea. Rare nationally, limited to SW England, and decreasing. All sites on dry sandy grassland, including one inland one. Another inland site is on disturbed sandy ground and two additional inland sites are on 'made up' land, but are not planted however. Recorded by Linton and Townsend as abundant around Christchurch Harbour and Hengistbury Head, but very few sites elsewhere. Other sites in Dorset are around Poole Harbour.

Ornithopus perpusillus L.
Bird's-foot

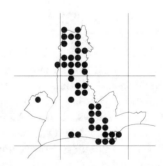

Fairly common; heaths and dry sandy places. Recorded from most areas of dry sandy grassland and heaths, but not generally found elsewhere.

ROSACEAE

Spiraea salicifolia L.
Bridewort

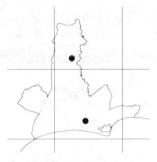

Introduced, uncommon; often planted and sometimes naturalized. One plant in Avon Forest Park South, in a hedgerow in 1985, and several by Iford Lane.

Filipendula ulmaria (L.)Maxim.
Meadowsweet

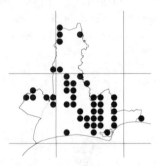

Fairly common; wet meadows, fens and marshes. A distinctly riverine distribution, recorded from the Avon and Stour valleys, and the Moors River and River Mude. One record in a wet flush on the cliffs at Honeycombe Chine.

Rubus idaeus L.
Raspberry

Infrequent; damp woods and heaths. Several records from damp woods and hedgerows by the River Stour, and five from heathland areas.

Rubus caesius L.
Dewberry

Uncommon; hedgerows, scrub and grassland. One record from grassland and hedgerows near Throop Mill, Bournemouth. Undoubtedly under-recorded now, previously recorded as locally abundant by Linton, although few records generally for Dorset.
1 grid square;19

Rubus fruticosus sens.lat.
Bramble

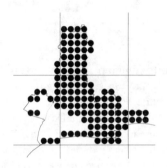

Very common everywhere, throughout the area. Recorded from every grid square.

Potentilla palustris (L.)Scop.
Marsh Cinquefoil

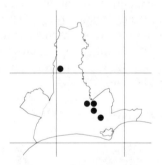

Infrequent; wet boggy places, fens, marshes and wet heaths. Only four records from the Avon valley, and one in the Moors River valley. Townsend and Linton recorded it from many sites, and as common along the Stour tributaries west of Hurn. Much decreased now, and very vulnerable to drainage operations.

Potentilla sterilis (L.)Garcke
Barren Strawberry

Infrequent; hedgebanks and woods on dry soils. Two records from grassy heathland, and three from more wooded areas. Recorded as very common by Linton and Townsend, much less frequent now.

Potentilla anserina L.
Silverweed

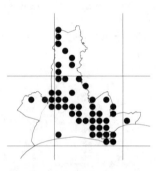

Fairly common; damp meadows, roadsides and waste places. Mainly recorded from the Stour and Avon valleys and around Christchurch Harbour. Much less common on the drier soils and heathland areas.

Potentilla recta L.
Sulphur Cinquefoil
Introduced, uncommon; a garden escape or casual, somtimes naturalized. A single site in Avon Forest Park in heathy grassland, growing by a small path through bracken. Recorded in 1985, and present most years since. Only recorded from a few sites in Dorset (Good, 1984), and not mentioned in old Floras.
1 grid square;10

Potentilla erecta (L.)Rauschel
Tormentil

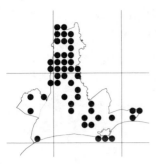

Fairly common; heaths and grassland on acid soils. Common on heaths and commons, and on dry acid grasslands in the north of the area, much less frequent elsewhere.

Potentilla anglica Laicharding
Trailing Tormentil

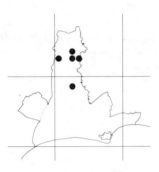

Infrequent; dry grassland and heaths. Several records, all from heathlands north of Hurn, and between the Moors River and River Avon. Recorded previously by Linton and Townsend from heaths in the south of the area.

Potentilla reptans L.
Creeping Cinquefoil

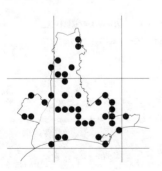

Frequent; hedgerows, roadsides and grassland. Records scattered widely throughout the area, but many noticeable gaps. Previously recorded by Linton and Townsend as very common except on the heathlands.

Fragaria vesca L.
Wild Strawberry

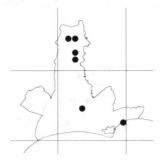

Infrequent; woods and hedgebanks on base-rich soils. One site on dry hedgebanks near Holdenhurst, one in woodland and the other sites in dry sandy grassland. Previously very common, especially off the heathlands.

*Fragaria x ananassa Duchesne
Garden Strawberry
Introduced, uncommon; garden escape. Plants growing on the cliffs between Bournemouth East Cliff and Toft zigzag, found in 1987.
1 grid square;19

Geum urbanum L.
Wood Avens

Frequent; woods, hedgebanks and damp shady places. Common in the Stour valley, and in the woodlands east of Christchurch. Very few records from the lighter heathland soils.

Geum rivale L.
Water Avens
Uncommon; wet marshy meadows. One site in wet meadows in the Avon valley, with a small colony. Only 12 plants flowered in 1989, the colony is therefore very vulnerable. Previously recorded as local in various flood meadows locally, but obviously this species has declined considerably.
1 grid square;19

Agrimonia eupatoria L.
Agrimony

Infrequent; roadsides, hedgebanks and field borders. Mainly recorded from hedgerows and field borders to the east of the River Avon, very few records elsewhere. Townsend and Linton recorded it as very common, and common off the heathland, certainly less frequent now.

Aphanes arvensis L. sens.lat.
Parsley-piert

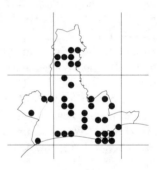

Frequent; arable land and dry soils. Recorded from most heathlands and dry sandy grasslands, especially along the coast, and in the north of the area.

Aphanes microcarpa(Bois & Reut.)Rothm.
Slender Parsley-piert

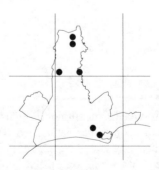

Infrequent; dry sandy grassland in short turf or sparse areas. All records from dry sandy areas, often in large colonies. Undoubtedly under-recorded as many records included in *A. arvensis* sens.lat. Not distinguished in Linton or Townsend.

Sanguisorba minor Scop. **subsp. minor**
Salad Burnet

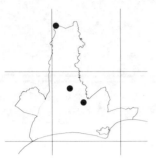

Uncommon; usually calcareous grassland, occasionally neutral grassland. Three records, all from sandy and gravelly areas, one in Moors Valley Country Park, one on Sopley Common and the other on a gravel mound on St Catherines Hill, all unlikely locations. Not recorded for this area in old Floras.

***Sanguisorba minor subsp. muricata** Briq.
Fodder Burnet
Introduced, uncommon; often naturalized on field borders and other cultivated places. A single plant growing on an old spoil heap excavated from a new waterbed at West Hants Water Company, Christchurch in 1982, now no longer there.
1 grid square;19

***Rosa rugosa** Thunb.
Japanese Rose

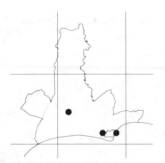

Introduced, uncommon; a garden escape, sometimes naturalized. Two plants, one with dark pink flowers, one with white, on Spellars Point, Stanpit Marsh, recorded in 1991, and one on Hengistbury Head in 1993. Also recorded on waste land in North Bournemouth, probably bird-sown. Described in Good (1984) as an occasional garden escape, likely to spread.

Rosa canina L. agg.
Dog-rose

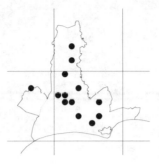

Infrequent; woods and hedgerows. Recorded from the Stour valley, with a few other scattered records. Undoubtedly under-recorded, with many records included in *Rosa* agg.

Rosa agg.
Wild Rose

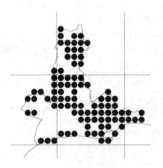

Very common; woods, hedgerows and scrub. Widespread throughout the area, probably present in most grid squares, further work is required on individual species' distribution. Linton and Townsend list 8 species for this area.

Prunus spinosa L.
Blackthorn

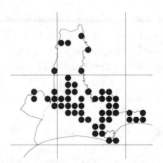

Fairly common; hedgerows, thickets and wood borders. Most frequent in the Stour valley and Moors River valley, and east of Burton and Winkton, these being the areas with the most hedgerows. Perhaps more frequent previously as Linton and Townsend describe it as very common, the

decline is presumably related to recent loss of hedgerows.

***Prunus domestica** L.
Wild Plum

Introduced, uncommon; hedgerows, usually near houses. Recorded from Riverside Lane, Iford in woodland and scrub by the river, and at Iford Bridge, known there since 1879 (Townsend).

Prunus domestica subsp. insititia
(L.)C.K.Schneider
Bullace

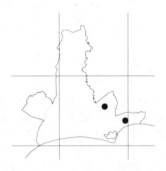

Possibly introduced, uncommon; hedgerows and thickets. Both records in hedgerows by green lanes. Many scattered sites listed in Linton and Townsend, probably under-recorded now.

***Prunus laurocerasus** L.
Cherry Laurel

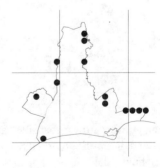

Introduced, infrequent; commonly planted in gardens, often self-sown and naturalized. Recorded from several widely spread sites around the area, mostly planted or self-sown from nearby plants.

Prunus lusitanica L.
Portugal Laurel

Introduced, uncommon; usually planted. All are records of planted shrubs, the one in Avon Forest Park is on the site of an old garden.

Cotoneaster simonsii + spp.
Cotoneaster

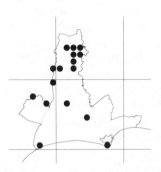

Introduced, infrequent; often planted and sometimes naturalized. Recorded mainly from heathland, especially around Avon Forest Park; some of these plants were probably originally planted, some may be bird-sown.

Cotoneaster franchetii Bois
Franchet's Cotoneaster
Introduced, uncommon; sometimes naturalized. One record in 1993 on the edge of forestry woodland at Ashley Heath.
1 grid square;10

Pyracantha coccinea Roem.
Firethorn
Introduced, uncommon; garden escape, sometimes naturalized. A record on rough grassland by the River Stour at Muscliff, and a tree-sized plant in the hedgerow of a lane at West Hurn.
2 grid squares;09,19

Crataegus monogyna Jacq.
Hawthorn

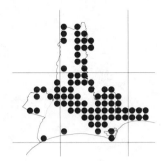

Very common; woods, scrub and hedgerows, often planted. Present in most areas, although not so frequent along the coast and on the sandy heathlands.

Amelanchier x lamarckii Schroeder
Juneberry

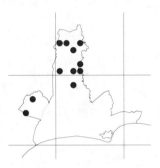

Introduced, infrequent; often grown in gardens and sometimes naturalized on light soils. Recorded mainly on the sandy heaths in the north of the area, particularly around Avon Forest Park.

Sorbus aucuparia L.
Rowan

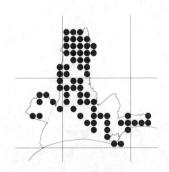

Common; woods and copses, often planted. Most frequent on the dry sandy soils and heathlands, and much less common elsewhere.

Sorbus aria sens.lat.
Whitebeam

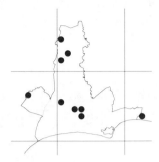

Infrequent; usually in hedgerows and thickets on chalk and limestone soils. Most sites are in woodland or scrub, all of them are probably either planted or grown from bird-sown berries.

Malus sylvestris sens.lat.
Crab Apple

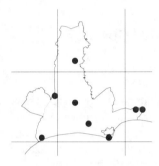

Infrequent; hedgerows and woods. Several records from the Stour valley, and two in Highcliffe. Previously recorded as not common in Hants (Townsend) and local around Bournemouth (Linton).

Malus domestica Borkh.
Apple
Introduced; escape from cultivation. A record from Avon Forest Park North, on the site of an old garden or smallholding, now open countryside.
1 grid square;10

PLATANACEAE

Platanus hybrida Brot.
London Plane

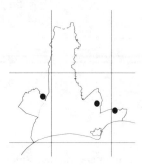

Introduced, uncommon; often planted and sometimes naturalized. All three sites on roadsides, two of them by woodland, probably all planted. Commonly planted in towns, but uncommon in rural areas.

CRASSULACEAE

Sedum telephium L.
Orpine
Uncommon; hedgebanks and woods. Recorded under pines on the edge of a plantation at Ashley Heath in 1993. Previously recorded by Linton and Townsend in this area only at Sopley.
1 grid square;10

Sedum anglicum Hudson
English Stonecrop

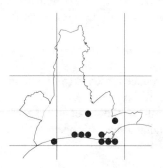

Infrequent; sandy places near the sea. Most sites are on sandy grassland or dunes by the sea where it is often frequent. Only two inland sites, one is on sandy grassland, the other on the concrete of a disused airfield.

Sedum album L.
White Stonecrop

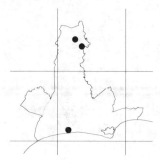

Introduced, uncommon; on walls and rocks. One record in Avon Forest Park on the edge of a concrete road, one beside a pine wood, the other from the cliffs at Boscombe. Not listed for this area in old Floras.

Sedum acre L.
Biting Stonecrop

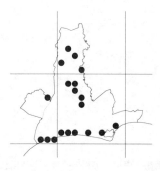

Infrequent; dry grassland, walls and sandy places. Recorded mainly along the coast, but also sometimes inland, on dry sandy grassland. Also recorded along the edges of some roads on sandy soils.

Crassula tillaea L.-Garland
Mossy Stonecrop

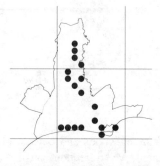

Rare; sandy and gravelly ground. Main centres in UK are East Anglia and Dorset, and declining nationally. Mainly found in bare sandy and gravelly places along the coast, but also several sites inland on sandy paths and tracks. Apparently increasing locally and some colonies quite

substantial, although not covering very large areas. Previously also recorded at Iford, Christchurch, Ensbury, Highcliff and Chewton Common (Linton, Townsend).

SAXIFRAGACEAE

Saxifraga x urbium D.A. Webb
Londonpride
Introduced, uncommon; commonly grown in gardens, sometimes naturalized. A few plants recorded in 1993 on the edge of a forestry plantation at Ashley Heath.
1 grid square;10

Saxifraga tridactylites L.
Rue-leaved Saxifrage
Uncommon; walls and dry places, usually on basic soils. One record from the bridge at Christchurch, seen in 1993. Previously recorded by Townsend also at Hengistbury.
1 grid square;19

Saxifraga granulata L.
Meadow Saxifrage
Uncommon; well-drained grassy places. A single site with a small colony of plants on a mossy grassy verge beside the drive of a large house. Not previously recorded in local Floras for this area, declining nationally.
1 grid square;10

Chrysosplenium oppositifolium L.
Opposite-leaved Golden-saxifrage
Uncommon; wet shady woods, streamsides and wet rocks. One site in the east of the area, in a wet shady woodland, growing on wet rocks by a stream. Linton and Townsend describe it as very rare in this area, and only record one site for this district.
1 grid square;29

Tellima grandiflora (Pursh)Douglas ex Lindley
Fringe-cups
Introduced, uncommon; sometimes naturalized in woods and damp shady places. A colony of a dozen plants recorded in 1993 beside a stream in a damp meadow in North Bournemouth. All petals in bud and flowers were a dark red-purple. Not previously recorded in old Floras for

this area. Only recorded in Dorset from one area of Purbeck.
1 grid square;09

ESCALLONIACEAE

Escallonia rubra (Ruiz & Pavon)Pers.
Escallonia
Introduced, uncommon; on cliffs near the sea. Two sites with several plants, on Bournemouth East Cliff and near Portman Ravine. Possibly planted, but often self-sown and naturalized.
2 grid squares;19

GROSSULARIACEAE

Ribes rubrum L. sens.str.
Red Currant

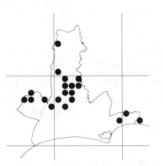

Frequent; woods, damp hedges and by streams. Almost all records from the Stour and Moors River valleys. Recorded by Linton as frequent in the Stour watershed. More common in southeast England than in the southwest.

Ribes nigrum L.
Black Currant

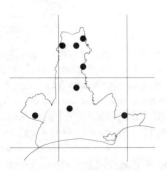

Infrequent; cultivated and an escape, but sometimes naturalized or possibly native in damp woods and copses. Three sites in damp deciduous woodland, and two in carr near rivers.

*Ribes sanguineum Pursh
Flowering Currant

Introduced, uncommon; garden escape, sometimes naturalized. One plant at Redhill, by the River Stour in rough grassland, probably planted, and some in Middle Chine, also probably planted.

Ribes uva-crispa L.
Gooseberry

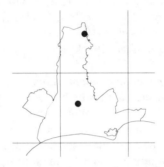

Uncommon; woods and copses, often an escape from cultivation, but sometimes probably native by streams in woods. One record from a copse by Holdenhurst village, possibly an escape from cultivation, and one near Ashley. Previously recorded by Linton at Kinson.

DROSERACEAE

Drosera rotundifolia L.
Round-leaved Sundew

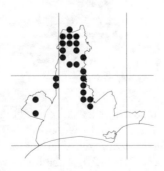

Frequent; wet heaths and bogs. Recorded on Kinson Common, Town Common and on most heaths north of Hurn.

Drosera intermedia Hayne
Oblong-leaved Sundew

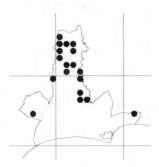

Infrequent; heaths and damp peaty places. Recorded mainly from heaths in the north of the area, generally in slightly wetter places than *D. rotundifolia*.

LYTHRACEAE

Lythrum salicaria L.
Purple-loosestrife

Frequent; riverbanks, marshes and ditches. Common by the River Stour, often recorded in the Avon valley, and by the Moors River and River Mude.

Lythrum portula (L.)D.A. Webb
Water-purslane

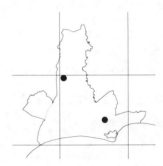

Uncommon; muddy pools and ponds. Several plants in 1983 at Purewell Meadows, now threatened by housing development; and a few plants in a dried up pond on East Parley Common in 1991. Previously common in the Avon valley, and recorded at Bournemouth and Hengistbury (Linton, Townsend).

ELEAGNACEAE

*Hippophae rhamnoides L.
Sea-buckthorn
Uncommon; introduced locally, planted on sand dunes. Several plants on Hengistbury Head sand spit, planted to help stabilise the sand dunes, recorded in 1986.
1 grid square;19

ONAGRACEAE

Epilobium hirsutum L.
Great Willowherb

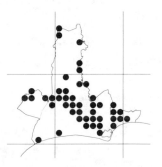

Fairly common; marshes, ditchsides and riverbanks. Very common in the Stour valley and Avon valley, also found by most other streams and ditches in the area.

Epilobium parviflorum Schreber
Hoary Willowherb
Uncommon; marshes, ditches and streambanks. A single site in marshy ground near the River Avon above Christchurch. Linton and Townsend recorded it as common and did not bother to give locations; it has therefore certainly decreased considerably since then.
1 grid square;19

Epilobium montanum L.
Broad-leaved Willowherb

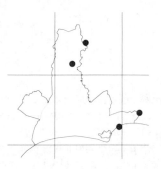

Infrequent; hedgerows and woods on the more base-rich soils. Probably

under-recorded as generally common throughout the British Isles, also as a weed in gardens.

Epilobium lanceolatum Sebastiani & Mauri
Spear-leaved Willowherb
Rare; roadsides, banks and walls. A single site on a dry path, under trees in Walkford in 1989, probably this species but not confirmed; further records are required.
1 grid square;29

Epilobium roseum Schreber
Pale Willowherb
Uncommon; damp places and cultivated ground. A single record in 1981 from a lane in Middle Bockhampton, recorded as "possibly *E. roseum*". Until refound and confirmed it cannot be accepted as anything more definite. Previously recorded near Throop, Holdenhurst and Christchurch (Linton, Townsend).
1 grid square;19

*Epilobium ciliatum Rafin.
American Willowherb
Introduced, uncommon; waste places and gardens. First recorded in England in 1891 in Leicestershire, and now spreading rapidly in southeast England. Recorded from waste ground in North Bournemouth in 1993. Probably under-recorded, but not previously recorded in this area by Linton or Townsend, although shown in the Avon valley in ABF, 1962.
1 grid square;19

Epilobium tetragonum L.
Square-stalked Willowherb

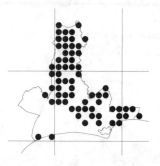

Common; hedgebanks, damp verges, ditchsides and woodland edges. Widely distributed over the whole area. Previous distribution is very uncertain due to confusion with *E. obscurum*. Unless glandular hairs or

stolons were definitely present, specimens were recorded under this species and not as *E. obscurum*.

Epilobium obscurum Schreber
Short-fruited Willowherb
Uncommon; marshes, streams, moist banks and woods. A single record from marshy ground near the River Avon at Winkton.
1 grid square;19

Epilobium palustre L.
Marsh Willowherb

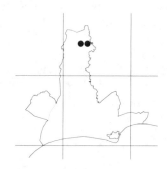

Uncommon; marshes and boggy places, a calcifuge. Two sites, both in damp grassy areas near Ashley Heath. Linton and Townsend also recorded this species from meadows in the Avon and Stour valleys.

Chamaenerion angustifolium (L.)Scop.
Rosebay Willowherb

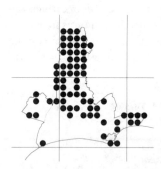

Common; wood clearings, burnt heathland, disturbed and waste ground. Recorded widely throughout the area, but most common on light soils in the north. Uncommon in 1900, Linton and Townsend listed only six sites around Bournemouth and Poole, but Rayner states it was greatly increased throughout Hampshire by 1929. Now vastly increased in distribution since 1900.

*Oenothera stricta Ledeb. ex Link
Fragrant Evening-primrose

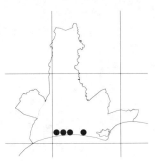

Introduced, very rare nationally; sand dunes and sandy places by the coast. Recorded in many places on the cliffs and clifftops from Bournemouth to Southbourne in 1992. Not listed in old Floras for this area.

*Oenothera biennis L.
Common Evening-primrose

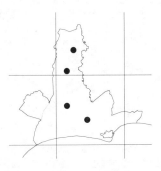

Introduced, infrequent; sandy ground and waste places. Two records on dry sandy heathy grassland, and the two Bournemouth records from a roadside verge and rough dry grassland. Recorded at Knapp Mill in 1924 (Rayner).

*Oenothera erythrosepala Borbas
Large-flowered Evening-primrose

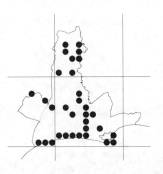

Introduced, frequent; roadsides, dunes, sandy cliffs and waste places. Recorded on the sandy cliffs along the coast, and in several places in the Stour valley. Also found on the roadside verges of the A338, and on the sandy heaths in the north of the

area. Not mentioned by Linton or Townsend, but now frequent and increasing and spreading widely.

Circaea lutetiana L.
Enchanter's-nightshade

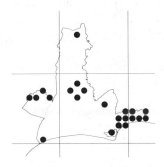

Frequent; woods and shady places on moist nutrient-rich soils. Recorded mainly from the Stour valley and in woods east of Christchurch. Although described as common by Linton and Townsend, only listed by them in Bournemouth, the Stour watershed and at Highcliff and Chewton Farm. Possibly never actually as common in this area as it has been described.

HALORAGINACEAE

Myriophyllum spicatum L.
Spiked Water-milfoil

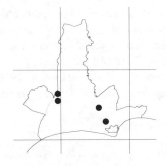

Infrequent; rivers, ponds and ditches. Recorded from both the lower Stour and Avon Rivers and in pools and ditches in the floodplains. Previous distribution of species in Dorset is uncertain, and records are few.

*Myriophyllum aquaticum
(Velloso)Verdcourt
Parrot's Feathers
Introduced, uncommon; ponds and pools. A single record from Wick Pond in 1989, introduced probably when the pond was replanted a few years before. Not mentioned in previous Floras.
1 grid square;19

HIPPURIDACEAE

Hippuris vulgaris L.
Mare's-tail
Rare; pools and slow streams, especially in base-rich water. One site on Stanpit Marsh, where it has been since before 1900 (Linton). The colony fluctuates from year to year according to the climate and grazing. A very small colony was also found on Priory Marsh in 1984. Local in the rest of Dorset, recorded from only six areas since 1980.
1 grid square;19

CALLITRICHACEAE

Callitriche stagnalis Scop.
Common Water-starwort

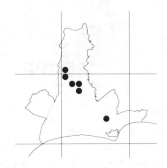

Infrequent; ditches, streams and wet mud. Several records from the Moors River, and one from mud in the lower Avon valley. Undoubtedly under-recorded, and some records included in *Callitriche* agg.

Callitriche agg.
Water-starwort

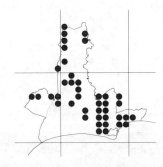

Fairly common; ponds, ditches and wet places. Many records in the Avon valley, and in the Stour and Moors River valleys. Distribution of individual species is uncertain.

CORNACEAE

Cornus sanguinea L.
Dogwood

Uncommon; usually in woods and hedges on calcareous soils. Several plants, presumably planted, along the centre of the old railway line at Ashley Heath. Recorded by Linton in hedges and woodland, and rather frequent off the heath.
1 grid square;10

ARALIACEAE

Hedera helix L.
Ivy

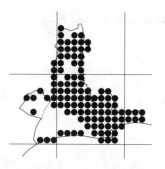

Very common everywhere; trees, hedges, in woods and on walls. Recorded throughout the area, in almost all grid squares.

UMBELLIFERAE

Hydrocotyle vulgaris L.
Marsh Pennywort

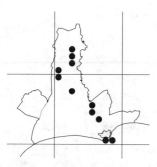

Infrequent; bogs, marshes and wet meadows. Recorded in the Avon valley and Moors River valley. Much less common than might be expected. Probably decreased, as Linton and Townsend recorded it as common.

Sanicula europaea L.
Sanicle

Uncommon; woods, especially beech woods and oak woods. A single site recorded in 1985 in a damp deciduous wood near Highcliffe. Previously recorded by Linton also at Christchurch.
1 grid square;29

Chaerophyllum temulentum L.
Rough Chervil

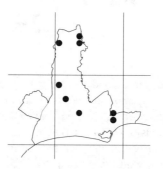

Infrequent; hedges, shady and grassy places. A few scattered records. Recorded by Linton, Townsend and Good as common, but not apparently, in this area now.

Anthriscus caucalis Bieb.
Bur Chervil

Rare; sandy ground and waste places. A single record in 1988 on Hengistbury sand spit. Obviously much declined as Townsend recorded it as frequent in the Stour watershed and at Mudeford, and Linton at Throop and at Purewell in plenty. Also recorded at Wick (Linton, 1919). Only seven sites in Dorset recorded since 1970.
1 grid square;19

Anthriscus sylvestris (L.)Hoffm.
Cow Parsley

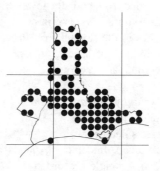

Very common; roadsides, hedgerows, wood margins, shady and waste places. Widespread throughout the area, but less common on the drier sandy soils especially in the north of the area.

*Smyrnium olusatrum L.
Alexanders

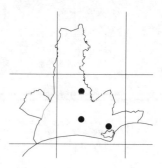

Introduced, uncommon; waste ground, hedges and cliffs. Two sites on roadside verges and one on rough ground beside a car park at Purewell. Recorded by Linton and Townsend at Iford, Christchurch and Purewell; and at Sopley (Linton, 1925).

Conopodium majus (Gouan)Loret
Pignut

Uncommon; woods and fields on acid soils. Two sites in scrub and grassland in the Stour valley, and an additional site at Dudsbury, also by the River Stour, but just in VC9. Recorded as common by Linton and Townsend. Probably under-recorded as this species is common throughout Britain except on chalk and in the fens.

Aegopodium podagraria L.
Ground-elder

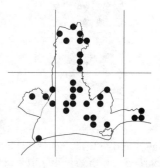

Frequent; disturbed ground and shady places, usually near houses, possibly introduced. Scattered throughout the area, often common near old buildings. Possibly originally introduced by the Romans, and later cultivated as a pot-herb.

Berula erecta (Hudson)Coville
Lesser Water-parsnip

Infrequent; ditches and streams. Several sites in the Avon valley, in ditches not far from the River Avon. No records from the lower Stour valley. Previously this species was recorded in the upper Stour valley (Mansel-Pleydell) however it was not recorded by Linton or Townsend from the lower Stour valley, only from meadows around Christchurch. Recorded later at Tuckton Creek (Linton, 1925).

Crithmum maritimum L.
Rock Samphire

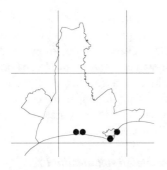

Infrequent; sea cliffs and sandy beaches. Numerous plants growing on the cliffs and on sandy beaches at several locations along the coast. Recorded by Townsend and Linton at Hengistbury and Mudeford, both sites now of course subject to over-use making survival very difficult.

Oenanthe fistulosa L.
Tubular Water-dropwort

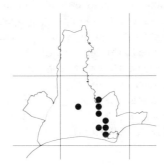

Infrequent; marshy places and ditches. Common in very wet meadows in the lower Avon valley, but only one record from the Stour valley. Linton and Townsend recorded it from the Avon meadows and fairly frequently along the River Stour and its tributaries, apparently much decreased now in the Stour valley.

Oenanthe pimpinelloides L.
Corky-fruited Water-dropwort

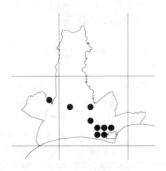

Infrequent; damp riverside meadows. Frequent in meadows in the Stour valley, and around Christchurch Harbour. Dorset, Hants and E. Devon are the main centres of the British population, elsewhere it is rare nationally. Possibly decreasing locally somewhat since 1900, as Linton and Townsend recorded it in many places, including on roadsides.

Oenanthe silaifolia Bieb.
Narrow-leaved Water-dropwort
Very rare; damp marshy meadows. One site on Stanpit Marsh, found in 1986 and 1987. Specimen confirmed by The Natural History Museum (London). Area is now more heavily grazed, and it has not been definitely refound so far. Rare nationally, only three other sites in Dorset, and only one site in Hants mentioned in Townsend.
1 grid square;19

Oenanthe lachenalii C.C.Gmelin
Parsley Water-dropwort

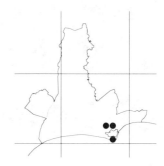

Uncommon; brackish marshes and grassland. Recorded in brackish grassland around Christchurch Harbour, and in places fairly frequent. Previously recorded by Linton and Townsend also at Mudeford.

Oenanthe crocata L.
Hemlock Water-dropwort

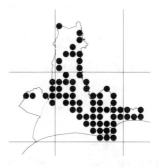

Common; streams, ditches and other wet places. Very common in the Stour and Avon valleys, and recorded from the Moors River, River Mude and other streams and ditches. The whole plant is very poisonous.

Oenanthe fluviatilis (Bab.)Coleman
River Water-dropwort
Uncommon; in streams and rivers. Three plants seen in the Moors River, 1991-3. Previously recorded by Linton and Townsend in the Rivers Stour and Avon. Much decreased since then.
1 grid square;19

Aethusa cynapium L.
Fool's Parsley

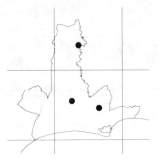

Uncommon; a weed of cultivated ground. Surprisingly only three records for this normally common and widely distributed weed. More records ought to be forthcoming.

Foeniculum vulgare Miller
Fennel

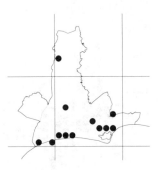

Infrequent; sea cliffs and waste ground, perhaps native, sometimes an escape. Mainly recorded by the coast, with some scattered records inland. Abundant on Bournemouth East Cliff in 1906 (Linton, 1919).

Conium maculatum L.
Hemlock

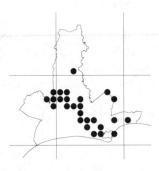

Frequent; riverbanks, hedges and damp places. Very common by the River Stour and nearby, but only a few other scattered records. Recorded by Townsend and Linton in the Stour valley, and around Christchurch.

Apium graveolens L.
Wild Celery
Uncommon; saltmarsh and brackish places. A few clumps on Stanpit Marsh. A decreasing species, described as frequent by Linton, who recorded it at Stanpit Marsh, Mudeford and about Hengistbury.
1 grid square;19

Apium nodiflorun (L.)Lag.
Fool's Water-cress

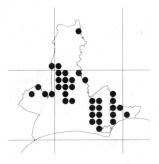

Frequent; ditches, streams and wet places. Recorded mainly from the Avon and Stour valleys, the Moors River and River Mude.

Apium inundatum (L.)Reichenb.fil.
Lesser Marshwort
Rare; shallow ponds and ditches. In ditches on Cowards Marsh, recorded for the first time in 1991, several plants. Also in 1991 (out of Dorset) across the River Avon near Sopley in larger amounts. Previously recorded at Winton by Linton and Townsend.
1 grid square;19

Carum carvi L.
Caraway
Uncommon; sometimes native, but generally an introduced casual. One site in long grass in a playing field, previously used as a paddock, next to a caravan park in North Bournemouth. Three sites mentioned in Linton: Bourne valley, Longfleet and by Knapp Mill, Christchurch.
1 grid square;19

Angelica sylvestris L.
Wild Angelica

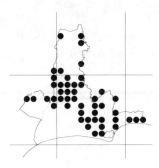

Fairly common; river and stream banks, marshes and wet places. Recorded in the Stour, Avon and Moors River valleys, and by the River Mude. Scattered records elsewhere, except on the heathlands.

Pastinaca sativa L.
Wild Parsnip

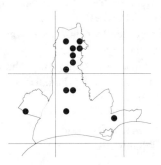

Infrequent; roadsides and grassy places, particularly on calcareous soils. Several records from the dry grasslands in and around Avon Forest Park, and a few isolated records. Generally common on the chalk, and not recorded for this area by Linton or Townsend.

Heracleum sphondylium L.
Hogweed

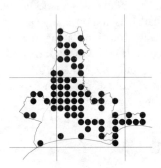

Common; hedgerows, roadsides and grassy places. Recorded widely throughout the area, but fewer records on the drier heathland soils, and along the coast.

Heracleum mantegazzianum
Sommier & Levier
Giant Hogweed

Introduced, infrequent; waste places, especially near rivers. Several records in the lower Avon valley, especially near the river around Wick. Also many plants recorded on Bournemouth cliffs. Not recorded for this area in previous Floras. This plant spreads aggressively and may become a nuisance, and can cause dermatitis, especially in strong sunlight.

Torilis japonica (Houtt.)DC.
Upright Hedge-parsley

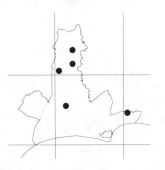

Infrequent; hedgebanks and grassy places. A few widely scattered records. Presumably under-recorded, as it is described in previous Floras as common and very common, but certainly not noticeably now in this area.

Daucus carota L.
Wild Carrot

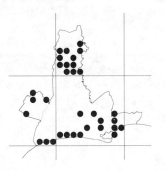

Frequent; grassy places and fields, especially near the sea. Recorded mainly on the sandy cliffs along the coast and on the dry sandy grasslands in the north of the area.

CUCURBITACEAE

Bryonia cretica subsp. dioica (Jacq.)
White Bryony

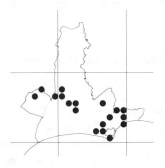

Infrequent; hedgerows and woodland. Recorded mainly in hedgerows in the Stour valley, and a few records from small woods to the east of Christchurch. Recorded by Linton from the Stour valley, Pokesdown, Hengistbury and Chewton Common.

EUPHORBIACEAE

Mercurialis perennis L.
Dog's Mercury
Uncommon; woods and shady places. One site only, a strong colony in Chewton Bunny. Recorded as common in woods and hedgebanks by Linton and Townsend. Some notable absences from this area, presumably has declined in recent years.
1 grid square;29

Mercurialis annua L.
Annual Mercury

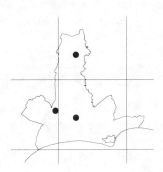

Uncommon; waste places and in cultivated ground, often a casual. Recorded in Avon Forest Park in 1985 growing in rough disturbed ground, as a weed in a flower bed in Muscliff in 1991, and also at Holdenhurst in 1992 by the roadside. Recorded by Townsend at Bournemouth in 1888 and Mudeford in 1889, also at Christchurch (Rayner).

Euphorbia lathyrus L.
Caper Spurge
Uncommon; probably introduced, a garden weed or escape. A single record in 1990 from rough ground at the edge of a car park near Ashley Heath. Possibly native in woods in a few places, but only previously recorded from this area and the rest of Dorset as a casual or escape (Linton; Good,1948).
1 grid square;10

Euphorbia helioscopia L.
Sun Spurge

Uncommon; mainly in cultivated ground. Two sites in disturbed ground, one in North Bournemouth, the other after regrading work on Bournemouth West Cliffs; and one near Iford Bridge in flowerbeds. Generally common in cultivated ground throughout the British Isles, probably much under-recorded locally.

Euphorbia peplus L.
Petty Spurge

Frequent; waste and cultivated ground, often a garden weed. Many records scattered throughout the area.

Euphorbia amygdaloides L.
Wood Spurge
Uncommon; generally in woods and thickets. A single record in 1992 on the edge of grassy heathland in Avon Forest Park. A very unusual habitat, presumably somehow introduced. Not previously recorded by Linton or Townsend for this area.
1 grid square;10

POLYGONACEAE

Polygonum aviculare L. agg.
Knotgrass

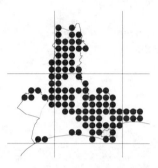

Very common everywhere; grows in most habitats. Probably present in every grid square.

Polygonum arenastrum Boreau
Equal-leaved Knotgrass

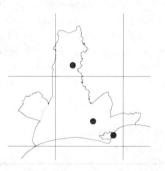

Uncommon; dry waste places and sandy ground. Two sites on dry shingly ground, and one on dry sandy grassland. Under-recorded although some records are included in the aggregate *P. aviculare*.

Polygonum oxyspermum subsp.raii
(Bab.)D.A.Webb & Chater
Ray's Knotgrass
Rare; on sandy shores. One site with several plants surviving. A nationally decreasing species, previously described as frequent and plentiful. Very few other sites on the south coast.
1 grid square

Polygonum maritimum L.
Sea Knotgrass
Extremely rare nationally; sandy shores. One site for this very rare species, with two well established plants and one smaller one in 1992, and 28 plants present in 1993. Described in CTM as extinct in England and local in the Channel Isles.
1 grid square

Polygonum bistorta L.
Common Bistort
Uncommon; damp meadows and grassland. Two sites, both by the River Stour. A few plants north of Bournemouth on grassland beside the river, and a small colony in long grass in meadows near Christchurch. Linton recorded it as rare and local but it was found at sites near Sturminster Marshall and higher up the River Stour. Not mentioned for this area in other previous Floras.
2 grid squares;19

Polygonum amphibium L.
Amphibious Bistort

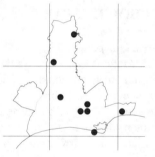

Infrequent; in rivers and ponds, and the terrestrial form in damp places. Present in the River Stour and its tributaries, and in the River Avon north of Christchurch. The terrestrial form recorded at Coward's Marsh, Hengistbury Head and near Ashley. Not as frequent now as recorded by Linton and Townsend.

Polygonum persicaria L.
Redshank

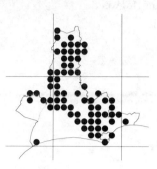

Common; cultivated ground, ditches and wet places. Recorded in the Stour and Avon valleys, by the River Mude and in the Moors River valley. Also found in damp places and ditches in the north of the area.

Polygonum lapathifolium L.
Pale Persicaria

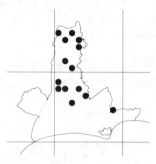

Infrequent; cultivated ground and damp places. Several scattered records, mainly in the north of the area. In similar places to *P.persicaria*, although not so frequent.

Polygonum hydropiper L.
Water-pepper

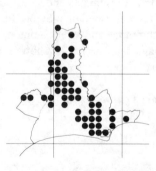

Fairly common; ditches, streamsides and wet places. Most common in the river valleys and on the heavier soils, and less frequent on the arable land and the drier sandy soils.

Fallopia convolvulus (L.)A Love
Black-bindweed

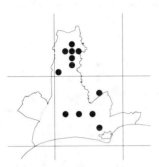

Infrequent; cultivated ground, hedges and waste places. Several scattered records but recorded mainly in and around Avon Forest Park. Described as common in Linton and Townsend, possibly under-recorded now.

*Fallopia baldschuanica
(Regel)J.Holub
Russian-vine
Introduced, uncommon; cultivated and often an outcast or escape. Recorded on the top of Boscombe Cliffs in 1993, as a patch a few metres across. Not mentioned in previous Floras of the area.
1 grid square;19

*Reynoutria japonica Hoult.
Japanese Knotweed

Introduced, infrequent; an escape from cultivation, often naturalized. Far too many widely scattered records of this fast-increasing invasive plant. Not mentioned in any of the old Floras.

Rumex acetosella agg.
Sheep's Sorrel

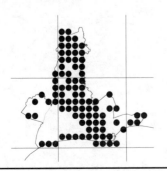

Very common; heaths, dry grassland and disturbed ground. Widespread throughout the area, although less frequent on the arable land east of Burton.

Rumex acetosa L.
Common Sorrel

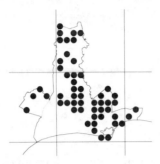

Fairly common; roadsides and grassy places, especially on heavier soils. Recorded mainly from the Avon, Stour and Moors River valleys and the River Mude. Much less frequent on the drier sandy soils and heathlands.

Rumex hydrolapathum Hudson
Water Dock

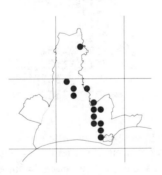

Infrequent; river banks, ditches and wet places. Recorded in the Avon valley, and in the Moors River valley. Not recorded by the River Stour, although Linton lists it in several places.

Rumex crispus L.
Curled Dock

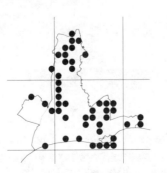

Fairly common; grassy places, waste and cultivated ground. Records scattered throughout the area,

probably much under-recorded as it is generally common throughout Britain.

Rumex obtusifolius L.
Broad-leaved Dock

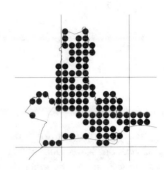

Very common everywhere; hedgerows, grassland and waste ground. Recorded in most grid squares, probably present in all of them.

Rumex sanguineus L.
Red-veined Dock

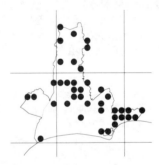

Frequent; grassland, hedges and in woods. Recorded in woodlands and hedgerows throughout the area, although nowhere very common, and fewer records in the north.

Rumex conglomeratus Murray
Clustered Dock

Frequent; roadsides, streamsides and damp places. Common in the Stour valley, also recorded in the Avon and Moors River valleys. Less frequent on the drier sandy soils.

Muehlenbeckia complexa
(A.Cunn.)Meissner
Wireplant

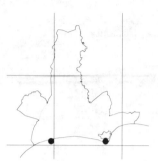

Introduced, uncommon; often naturalized on walls and cliffs. A straggling colony at the top of East Cliff, Bournemouth recorded in 1990. Also growing on Hengistbury Head in the old nursery area. A native of New Zealand, found in the Channel Isles, Scilly Isles and SW England. Not mentioned in previous Floras.

URTICACEAE

Parietaria judaica L.
Pellitory-of-the-wall

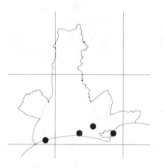

Infrequent; old walls, rocks and hedgebanks. Recorded in 1987 from the cliffs at Fishermans Walk, Southbourne, at West Cliff in 1992 and also several plants around the buildings at Mudeford Quay, and in 1993 on Christchurch Castle. Linton and Townsend described it as local and common on walls and bridges at several locations. Probably now under-recorded, although many plants are lost unnecessarily from old walls and bridges by over-zealous 'cleaning' operations.

Soleirolia soleirolii (Req.)Dandy
Mind-your-own-business
Introduced, uncommon; often planted in gardens and sometimes naturalized. Recorded in 1993 growing on the walls of Christchurch Castle keep.

Not mentioned by Linton or Townsend, but recorded in this area before 1962 (ABF).
1 grid square;19

Urtica urens L.
Small Nettle

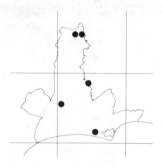

Infrequent; cultivated ground and waste places. Two sites on verges by cultivated ground, one beside a field and one on sandy ground.

Urtica dioica L.
Common Nettle

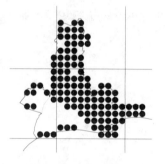

Very common everywhere; present in most habitats. Present in almost every grid square, and abundant in some.

CANNABACEAE

Humulus lupulus L.
Hop

Frequent; hedgerows and thickets. Common in hedgerows in the Stour valley, also recorded in the Avon valley and near the River Mude.

ULMACEAE

Ulmus minor Miller (incl. *U. procera*)
Elm

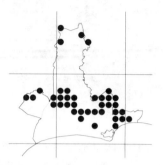

Frequent; hedgerows and roadsides, often planted, probably introduced. Recorded mainly in the Stour valley, and to the east of the River Avon. Very uncommon on heathland soils.

MYRICACEAE

Myrica gale L.
Bog-myrtle

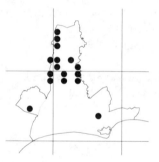

Infrequent; wet heaths and bogs. Mostly recorded on heaths and bogs north of Hurn and in the Moors River valley. Although found here, and in the Poole Basin, it is very uncommon in the rest of Dorset. Described by Linton as "abundant in many places in this district".

BETULACEAE

Betula pendula Roth
Silver Birch

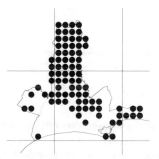

Very common; heathland, and in woods. Widespread throughout the area, but less common on the heavier soils and on arable land east of Burton. Not recorded along Bournemouth Cliffs.

Betula pubescens Ehrh.
Downy Birch

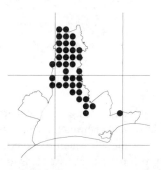

Fairly common; wet heaths and woods. Much less common than *B. pendula*, recorded mainly on the heathlands, especially in the north of the area.

Alnus glutinosa (L.)Gaertner
Alder

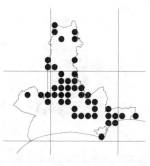

Fairly common; by streams and rivers, in marshes and wet places. Recorded in the Avon, Stour and Moors River valleys and by the River Mude, also by ditches and other wet places.

CORYLACEAE

Carpinus betulus L.
Hornbeam

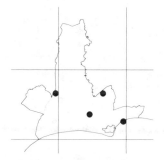

Infrequent; woods and hedgerows. Two sites definitely planted, the Steamer Point trees also possibly planted. This species is only native in southeast England, and when found elsewhere it has usually been planted.

Corylus avellana L.
Hazel

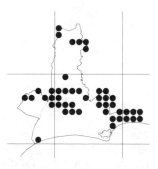

Fairly common; woods, copses and hedgerows. Recorded mainly in the Stour valley, and to the east of the River Avon. Although recorded by Linton and Townsend as generally common, Townsend did not record any locations in this area, and Linton only at Highcliff.

FAGACEAE

Fagus sylvatica L.
Beech

Frequent; woods and hedges, especially on drier soils, often planted. Many trees around Highcliffe, and on heathland soils in and around Avon Forest Park and Ashley Heath. Also recorded in the Stour valley. Many of these trees may have been planted, but self-seeding certainly occurs on the drier soils.

*Castanea sativa Miller
Sweet Chestnut

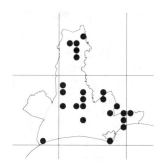

Introduced, frequent; often planted, sometimes naturalized. Scattered records throughout the area, mostly probably planted.

*Quercus cerris L.
Turkey Oak

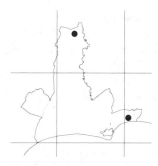

Introduced, uncommon; often planted and sometimes naturalized. Several trees recorded growing along the edge of the old Ashley Heath railway line and many at Highcliff Castle. Not previously recorded by Linton or Townsend for this area.

*Quercus ilex L.
Evergreen Oak

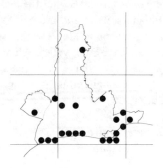

Introduced, frequent; often planted, sometimes naturalized. Common on the cliffs along the coast, a few scattered records inland.

Quercus robur L.
Pedunculate Oak

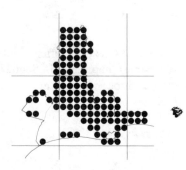

Very common everywhere; woods and hedgerows. Recorded from almost all grid squares, but not from parts of Bournemouth Cliffs.

*Quercus borealis Michx.
Red Oak

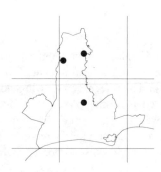

Introduced, uncommon; planted for ornament and forestry. All sites not very far from forestry areas, one record from the edge of a coniferous plantation.

SALICACEAE

*Populus alba L.
White Poplar

Introduced, infrequent; streambanks and wet places, usually planted. Several records on Bournemouth Cliffs, also in the Avon and Stour valleys.

Populus tremula L.
Aspen

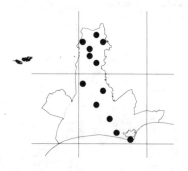

Infrequent; woods, and on poorer soils. A few scattered records, mostly on heathland in the northern half of the area.

*Populus x canescens (Aiton)Sm.
Grey Poplar

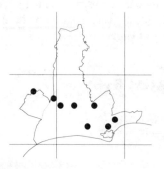

Introduced, infrequent; damp woods and hedges, often planted. Mainly recorded in the Stour valley, as noticed by Townsend.

*Populus nigra var italica Moench
Lombardy Poplar

Introduced, infrequent; usually planted. Four records, all on rough ground, three of them near the River Stour.

*Populus x canadensis Moench
Italian Poplar

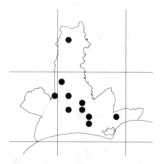

Introduced, infrequent; generally widely planted throughout the British Isles. Although usually planted, the few records are almost all from the Stour valley. Not mentioned in Linton or Townsend; Rayner recorded it at Lymington.

Salix fragilis L.
Crack Willow

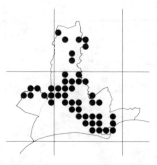

Fairly common; marshes, rivers and streams, often planted. Mainly recorded from the Stour valley, but also by the Avon and the River Mude.

Cuscuta epithymum
Dodder

PLATE 1.

Ophioglossum azoricum
Small Adder's-tongue

Parentucellia viscosa
Yellow Bartsia

Dianthus armeria
Deptford Pink

Spiranthes spiralis
Autumn Lady's-tresses

PLATE 2.

Orobanche minor
Common Broomrape

Orchis morio
Green-winged Orchid

Dactylorhiza maculata subsp. ericetorum
Heath Spotted-orchid

Limonium vulgare
Common Sea-lavender

PLATE 3.

Oenothera stricta
Fragrant Evening-primrose

Althaea officinalis
Marsh-mallow

Butomus umbellatus
Flowering-rush

Lychnis flos-cuculi
Ragged-Robin

PLATE 4.

Geum rivale
Water Avens

Potentilla palustris
Marsh Cinquefoil

Menyanthes trifoliata
Bogbean

Salix alba L.
White Willow
Uncommon; wet woods and by streams and rivers. A record from Merritown in 1984, from a wet wood by a tributary of the River Stour. Recorded from the Stour and Avon valleys by Linton and Townsend. Very few records from the western half of Dorset.
1 grid square;19

***Salix x sepulcralis** Simonkai
Weeping Willow
Introduced, uncommon; but widely cultivated and often planted. Mature trees by the River Stour at Muscliff, probably planted.
1 grid square;09

Salix triandra L.
Almond Willow

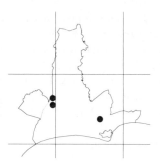

Uncommon; banks of rivers and streams, planted in osier beds. Two records from banks of the River Stour at Muscliff, and one from the Avon valley north of Christchurch. Probably under-recorded, Linton lists many sites in this area, but there are few other previous records for Dorset.

Salix purpurea L.
Purple Willow
Uncommon; by rivers and streams, sometimes planted. A single record from the banks of the River Stour at Throop Mill, in 1982. Probably under-recorded, previously found in the Avon valley by Linton, however not many records exist for Dorset.
1 grid square;19

Salix viminalis L.
Osier

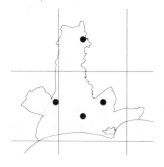

Uncommon; by streams and rivers, often planted. Two records from the banks of the Stour, and two near the River Avon. Probably under-recorded, a few sites listed by Linton, few records in the rest of Dorset.

Salix x mollissima Hoffm. ex Elwert
Sharp-stipuled Willow

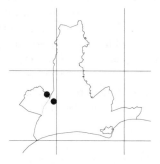

Uncommon; by rivers and streams, planted as an osier, or an escape. Two records, both from the banks of the River Stour at Muscliff, Bournemouth.

***Salix elaeagnos** Scop.
Olive Willow
Introduced; cultivated or garden escape, uncommon. Several bushes recorded in 1989 by the River Stour at Redhill, planted.
1 grid square;09

Salix caprea L.
Goat Willow

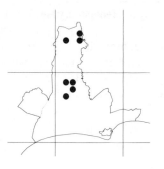

Uncommon; woods and hedges. Four sites near the Moors River in wet meadows and edges of damp woodland, and three sites around Ashley Heath. Recorded by Linton and Townsend as frequent and common, but not as common as *S. cinerea*. Possibly under-recorded, but still not widespread in this area.

Salix cinerea L. sens.lat.
Grey Willow

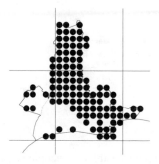

Very common everywhere; hedges, woods and heaths. Recorded in almost all grid squares.

Salix cinerea subsp. oleifolia
Macreight
Rusty Willow

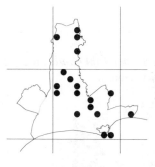

Infrequent; hedgerows, marshes and by streams. Widely scattered records, much under-represented as many records are included in *S. cinerea* sens.lat.

Salix x reichardtii A.Kerner
Uncommon; a hybrid of *S. caprea x cinerea subsp. oleifolia*. A single record at Avon Forest Park in 1985, growing amongst other sallow species on the edge of a damp grassy clearing. An unusual shrub as the catkins are androgynous. Not previously recorded in any Floras for this area.
1 grid square;10

Salix x smithiana Willd.
Silky-leaved Osier
Uncommon; damp places. A single
record in 1985 from Matchams View
in Avon Forest Park. Several previous
records in Linton and Townsend,
mainly from the Avon valley.
1 grid square;10

Salix repens L.
Creeping Willow

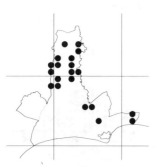

Frequent; heathland. Recorded mostly
on and near Barnsfield Heath and
around St Leonards. Only a few
records on heaths further south.
Previously recorded by Linton and
Townsend as common on the heaths
about Bournemouth and north of
Christchurch.

ERICACEAE

*Rhododendron ponticum L.
Rhododendron

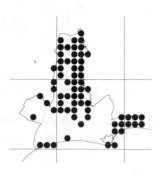

Introduced, common; originally
planted, but now completely
naturalized. Unfortunately common
and spreading on most of the heathland
areas and in some woods, and
becoming a nuisance in many areas
by ousting other plants. Not
mentioned in Linton or Townsend,
but praised for its splendid growth
around Heron Court (Hurn Court) in
1914 (Morris), and on the steep slope
of St Catherine's Hill described as
"the beautiful rhododendron woods

of Heron Court" (Walls, E. 1929).
The "Rhododendron forests" around
Hurn Court were apparently planted
by James Edward, the second Earl
of Malmesbury between 1796 and
1820 (Aflalo, 1905). An invasive
species which is increasing
relentlessly.

*Rhododendron - red
Introduced, uncommon; planted,
garden escape or occasionally
naturalized. Two records in Avon
Forest Park, both on the sites of old
gardens, and a record from Steamer
Point woodland, originally planted in
the extensive grounds of Highcliffe
Castle.
 2 grid squares;10,19

*Rhododendron luteum Sweet
Yellow Azalea
Introduced, uncommon; sometimes
cultivated and occasionally
naturalized. Three or four plants
growing well in Avon Forest Park
South amongst bracken, brambles
and *R. ponticum*. Not previously
recorded for this area, but increasing
on acid soils in Dorset.
1 grid square;10

*Gaultheria shallon Pursh
Shallon

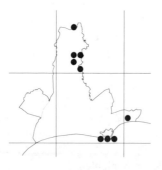

Introduced, often planted, and
sometimes naturalized. Recorded in
Avon Forest Park and near
Matchams, and on Hengistbury Head
and at Steamer Point. First recorded
locally on Hengistbury Head in 1962
(BNSS). An invasive species which
can become dense if left. Not
mentioned in old Floras, but now
increasing steadily and locally
dominant in some places.

*Arbutus unedo L.
Strawberry-tree
Introduced; native only in Ireland. A
single tree planted in woodland, in
grounds previously belonging to
Highcliffe Castle.
1 grid square;29

*Vaccinium macrocarpon Aiton
American Cranberry
Introduced, uncommon; sometimes
naturalized in peaty places. Two sites
recorded in very wet heathland near
Ashley Heath. One site is extensive,
and both are near the same stream but
about 0.5 km apart. Both sites are
additional to the site listed in Good
(1984). Not mentioned in other local
Floras for this area.
2 grid squares;10

*Vaccinium corymbosum L.
Blueberry
Introduced, uncommon; occasionally
naturalized on heathland. A large
mature shrub, recorded in 1993, in a
damp heathy woodland beside a
stream near Ashley Heath. Not
previously mentioned in local Floras
for this area, although listed by Stace
as naturalized in S. Hants. and Dorset.
1 grid square;10

Calluna vulgaris (L.)Hull
Ling, Heather

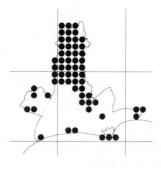

Common; heaths and commons.
Abundant on the sandy heathland
soils throughout the area, and in
places along the coast. On the coast
it seems to be commoner on the cliffs
where re-grading works have not
been carried out.

Erica tetralix L.
Cross-leaved Heath

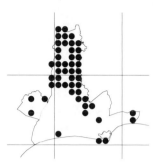

Fairly common; damp and wet heaths. Recorded from the wetter parts of most heathlands in the area. Not common on the cliffs by the coast, only where wet flushes occur.

Erica cinerea L.
Bell Heather

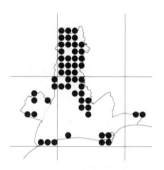

Fairly common; dry heaths. Present on the drier parts of almost all the heathlands in the area, and in many places along the cliffs. More common than E. tetralix.

*Erica lusitanica Rudolphi
Portugese Heather
Introduced, uncommon; sometimes cultivated and naturalized in Dorset and Cornwall. A single site on heathland amongst *Erica cinerea*, apparently naturalized. One other site in Dorset, at Lytchett Matravers, introduced in 1876 (Good, 1948).
1 grid square;10

PLUMBAGINACEAE

Limonium vulgare Miller
Common Sea-lavender

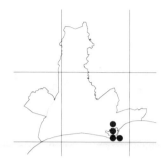

Infrequent; mudflats and saltmarshes. Common on the saltmarshes in Christchurch Harbour.

Armeria maritima (Miller)Willd.
Thrift

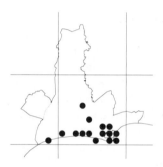

Infrequent; saltmarshes, cliffs and shores. Common around Christchurch Harbour, and recorded along the cliffs to Durley Chine. Two inland records, one on "tipped" land at Tuckton, the other recorded in 1989 on the central reservation of the A338, growing with *Cochlearia danica*, presumably because of the high salt content from road gritting.

*Armeria pseudarmeria
(Murray)Mansfield
Estoril Thrift

Introduced, uncommon; a garden escape. Two sites on Bournemouth cliff top to the east of the pier, found in 1989 and 1990; and one site on West Cliff. This species is much taller than *Armeria maritima* (about 30cm) and has flat leaves. It appears to have become naturalized here now.

Armeria maritima

PRIMULACEAE

Primula veris L.
Cowslip

Uncommon; old meadows and pastures. Several hundred plants in two old river meadows. Only recorded by Linton and Townsend from one location in this area, but an additional cowslip meadow in the Avon valley is mentioned in Walls, E.(1929).
1 grid square;19

Primula vulgaris Hudson
Primrose

Infrequent; woods and shady hedgebanks. Four sites in the area, all in deciduous woodland, although none with large numbers of plants present. Previously common, also in hedgebanks and field borders, but obviously declined considerably since 1900.
4 grid squares;09,10,19,29

Lysimachia nemorum L.
Yellow Pimpernel

Uncommon; moist woods and shady hedgebanks. One site on Stanpit Marsh in damp grassy scrubland. Only two sites listed in Linton and Townsend for this area.
1 grid square;19

Lysimachia nummularia L.
Creeping-Jenny

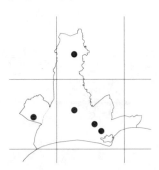

Infrequent; damp meadows and hedgebanks. Two records in damp grassy meadows and scrubland near Christchurch, one in meadows near Hurn, also on Turbary Common and at Avon Forest Park. Previously recorded by Linton and Townsend as common in the Stour valley, but now much less common.

Lysimachia vulgaris L.
Yellow Loosestrife

Infrequent; riverbanks, ditches and wet meadows. A few records in each of the Avon, Stour and Moors River valleys and by the River Mude, but nowhere common. Recorded by Linton and Townsend as common in the Stour valley, but certainly not now.

Anagallis tenella (L.)L.
Bog Pimpernel

Uncommon; damp grassy places and bogs. Several good patches in Purewell Meadows in 1982, probably still surviving but much overgrown; and a few small plants on East Parley Common. Much less common locally now, due probably to destruction of its habitat.
2 grid squares;19

Anagallis arvensis L.
Scarlet Pimpernel

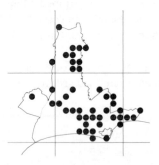

Fairly common; cultivated and waste ground and by roadsides. Many scattered records, mainly from roadsides and disturbed ground. The blue form (subsp. caerulea) was recorded in Pokesdown in 1903 (Townsend), and at Christchurch (Linton, 1925).

Glaux maritima L.
Sea-milkwort

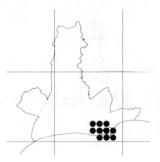

Infrequent; saltmarshes, seashores and estuaries. Only recorded in and around Christchurch Harbour, on saltmarshes and seashores.

Samolus valerandi L.
Brookweed

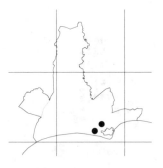

Uncommon; marshy places near the sea. Found on both sides of Christchurch Harbour, on Stanpit Marsh and Wick Meadows. Recorded in Linton and Townsend from several local sites, not now as common.

BUDDLEJACEAE

*Buddleja davidii Franchet
Butterfly-bush

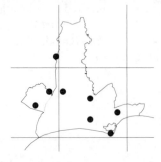

Introduced, infrequent; commonly grown in gardens, often naturalized. A few scattered records from waste places, by rivers and also on heathland edges. First introduced to England around 1890, but not mentioned in

Linton or Townsend. Recorded near Southampton before 1930, and since then spreading considerably.

OLEACEAE

Fraxinus excelsior L.
Ash

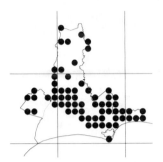

Common; woods and hedgerows, often planted. Recorded mainly in the Stour and Avon valleys, and to the east of the River Avon. Less common on sandy soils.

Syringa vulgaris L.
Lilac

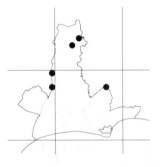

Introduced, infrequent; an escape or sometimes naturalized. Two records in hedgerows with several yards of the species, two by tracks on heathland and one record in Avon Forest Park on the site of an old garden, now grassland.

Ligustrum vulgare L.
Wild Privet

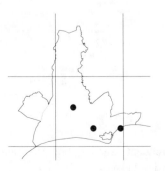

Infrequent; hedges and scrub, usually on calcareous soils. Three recorded sites only, but records before 1991 were recorded as *Ligustrum* agg. and have been included with *L. ovalifolium*. Previously recorded as frequent in woods and hedges by Linton and Townsend.

Ligustrum ovalifolium Hassk.
Garden Privet

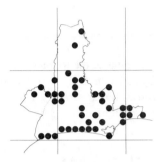

Introduced, fairly common; much planted in gardens and sometimes escaping. Many widely scattered records, some of which are most likely *L. vulgare*. Recorded frequently along the coast.

Forsythia x intermedia Hort ex Zobel
Forsythia

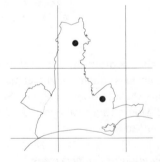

Introduced, uncommon; a garden escape. Plants in Avon Forest Park are remnants of previously gardened areas, the record from Winkton is a garden escape.

APOCYNACEAE

Vinca minor L.
Lesser Periwinkle

Infrequent; woods and hedgebanks. Several sites in hedgerows and woodland edges. Only listed previously in the Avon watershed (Linton, Townsend), and not very common in the rest of Dorset.

Vinca major L.
Greater Periwinkle

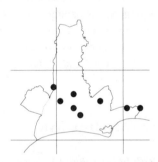

Introduced, infrequent; an escape and sometimes naturalized. A few scattered records, mostly escapes.

GENTIANACEAE

Centaurium pulchellum (Swartz)Druce
Lesser Centaury

Uncommon; damp grassy places near the sea. Several plants scattered in dense upper-saltmarsh turf at Stanpit Marsh, Christchurch, and Hengistbury Head. Described in Linton as frequent, but now much less common.

Centaurium erythraea Rafn
Common Centaury

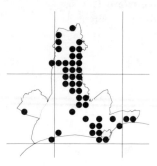

Fairly common; dry grassland and sandy places. Mainly recorded on the drier sandy soils, especially in the north of the area, and also in places along the coast.

Gentiana pneumonanthe L.
Marsh Gentian
Rare; wet heaths. The main areas in the British Isles for this species are the New Forest and Dorset. A species still decreasing, due mainly to loss or drying out of habitat, but also to unscrupulous collectors. Two sites with a few plants, and a further site in VC9 with a good population.
2 grid squares;19

MENYANTHACEAE

Menyanthes trifoliata L.
Bogbean

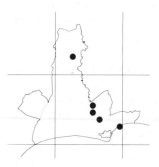

Infrequent; pools, wet meadows and boggy places. Good colonies at all three sites in the Avon valley. The Avon Forest Park plants were introduced into the new pond in 1991, it has also been planted in the pond at Steamer Point. Recorded by Linton and Townsend as common in the Stour valley and by the Bourne and Moors River. Considerably declined, due mainly to increased drainage.

Nymphoides peltata
(S.G.Gmelin)O.Kuntze
Fringed Water-lily

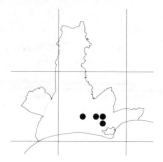

Rare; ponds and slow rivers. Three sites in and near the River Avon, and one by the Stour. Only two other sites in Dorset, but apparently increasing in distribution.

BORAGINACEAE

Symphytum officinale L.
Common Comfrey

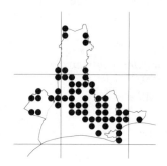

Fairly common; riverbanks, streamsides, ditches and wet places. Many records from the Avon, Stour and Moors River valleys and by the River Mude.

*Pentaglottis sempervirens
(L.)Tausch
Green Alkanet

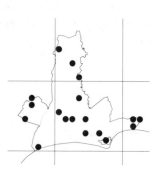

Introduced, infrequent; hedgerows and waste ground, sometimes naturalized. Several records from the Stour valley, and a few widely

scattered additional ones. Listed by Townsend at West Cliff, Bournemouth in 1889, not listed by Linton. Recorded at Southbourne in 1920, and Christchurch (Rayner), apparently increasing.

Anchusa arvensis(L.)Bieb.
Bugloss

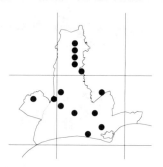

Infrequent; heaths, cultivated ground and sandy places. Several records from the Stour valley, and in and around Avon Forest Park. Recorded by Linton and Townsend as common and not unfrequent, so perhaps declined since.

Myosotis scorpioides L.
Water Forget-me-not

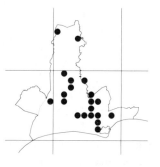

Frequent; streams, ditches and wet places. Recorded mainly in the Avon valley, and the Moors River valley.

Myosotis secunda A.Murray
Creeping Forget-me-not

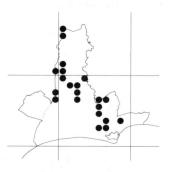

Frequent; ditches, boggy meadows and wet places. Fairly common in the Avon and Moors River valleys, but not frequent elsewhere.

Myosotis laxa Lehm*. **subsp.**
caespitosa(CFSchultz)Hyl.ex Nordh.
Tufted Forget-me-not

Uncommon; marshes, ditches and wet
meadows. Recorded in wet and
damp meadows in the Avon valley.
Previously recorded as common
throughout the area, possibly partly
under-recorded now.

Myosotis sylvatica Hoffm.
Wood Forget-me-not

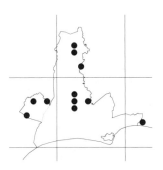

Infrequent; native only in damp
woods, otherwise a garden escape.
All records are of the common garden
species, and are escapes or throwouts,
a few of these may be naturalized.

Myosotis arvensis (L.)Hill
Field Forget-me-not

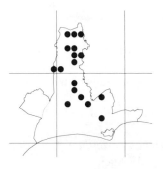

Infrequent; cultivated ground,
hedgerows and woods. Records
scattered throughout the area, in a
variety of habitats.

Myosotis discolor Pers.
Changing Forget-me-not

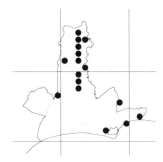

Infrequent; banks and dry grassy
places. Several sites on dry sandy
grassy banks, especially in Avon
Forest Park. Described by Townsend
as very common in the Stour
watershed.

Myosotis ramosissima Rochel
Early Forget-me-not

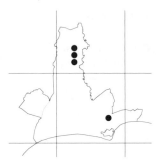

Infrequent; dry banks and sandy soils.
Three sites in Avon Forest Park, on
dry sandy grassland. The Purewell
Meadows site found in 1983, on a dry
grassy bank by Burton Road has
since been destroyed by housing.
Linton and Townsend recorded it
occasionally in the area, but there are
no recent records from the Stour
district where Townsend recorded it.

CONVOLVULACEAE

Convolvulus arvensis L.
Field Bindweed

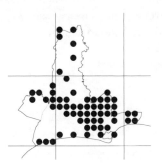

Common; cultivated and waste
ground, roadsides and hedgerows.
Recorded mainly in the Stour and
Avon valleys, and in hedgerows in
the arable area to the east of Burton.
Much less common in the north of the
area, especially on the heathlands.

Calystegia sepium (L.)R.Br.
Hedge Bindweed

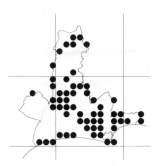

Fairly common; hedges, bushy
places, thickets and on riverbanks.
Recorded in the Avon, Stour and
Moors River valleys, by the River
Mude and along the coast. Not so
frequent on heathland soils.

Calystegia soldanella (L.)R.Br.
Sea Bindweed

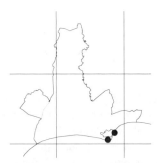

Rare; sandy and shingly shores and
dunes. Several plants recorded in
1990 on the dunes at Hengistbury
Head and also in 1993 on the sand
spit. Not found at Mudeford as listed
in Linton, seems to be declining in
Dorset as there are only seven records
since 1980.

Cuscuta epithymum (L.)L.
Dodder

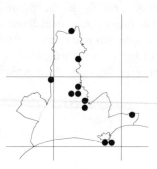

Infrequent; parasitic on gorse and heathers, especially *Calluna*. Several widely scattered records from heathland sites, sometimes present in large amounts. Not recorded from all heathland areas however.

SOLANACEAE

***Lycium barbarum** L.
Duke of Argyll's Teaplant

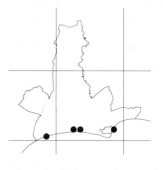

Introduced, infrequent; often naturalized in hedges and on walls, especially near the sea. Several records on the cliffs from Bournemouth to Southbourne, and at Hengistbury Head. Records in old Floras are sometimes confused; some combine *L. barbarum* and *L. chinense* into one species.

Solanum dulcamara L.
Bittersweet, Woody Nightshade

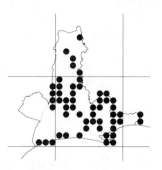

Fairly common; woods, hedgerows and riverbanks. Recorded from hedgerows and riverbanks in the

Stour, Moors River and Avon valleys, also from many locations along the cliffs and on shores around Hengistbury Head.

Solanum nigrum L.
Black Nightshade

Frequent; waste and cultivated ground. Records scattered throughout the area, but more common on the drier sandy soils, and around Throop.

***Solanum tuberosum** L.
Potato
Introduced, uncommon; an escape on waste and disturbed ground. One record in Avon Forest Park North, and one by the River Stour near Iford.
2 grid squares;10,19

***Lycopersicon esculentum** Mill.
Tomato
Introduced; escape from cultivation. A single plant, with flowers, growing on a muddy bank by the River Stour at Throop in 1984.
1 grid square;19

***Nicotiana alata** Link & Otto
Sweet Tobacco
Introduced, uncommon; often a casual or garden escape. A single record near Berry Hill, Throop, by a footpath. Probably a garden outcast.
1 grid square;19

***Petunia x hybrida** Vilm
Petunia
Introduced, garden escapes or casual. Several plants on dry sandy grass and heath at Matchams Stadium in 1986.
1 grid square;10

SCROPHULARIACEAE

Verbascum thapsus L.
Great Mullein

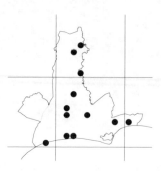

Infrequent; waste places, banks and hedgerows. A few records scattered throughout the area, probably under-recorded. Recorded by Linton as frequent, but less so now.

Verbascum nigrum L.
Dark Mullein

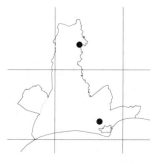

Uncommon; banks, roadsides and waste ground. Both records on areas of disturbance, one by a 'tipped' area on heath, the other by a recently laid track, probably of calcareous material. Very few previous records locally, recorded on waste ground at Christchurch (Linton, 1925).

***Antirrhinum majus** L.
Snapdragon
Introduced, uncommon; old walls and as a garden escape. One record from the disused railway track at Ashley Heath, probably a garden escape.
1 grid square;10

***Misopates orontium** (L.)Rafin.
Weasel's Snout
Uncommon; cultivated ground and sandy fields. A single record from dry sandy grassland on Kinson Common in 1986. Previously very common in many sandy fields throughout the area, and abundant near Christchurch (Linton). Has decreased considerably and now quite uncommon.
1 grid square;09

*Linaria purpurea (L.)Miller
Purple Toadflax

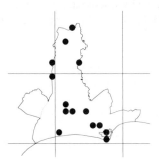

Introduced, infrequent; cultivated in gardens, sometimes naturalized. Several records in the Stour valley, and other scattered ones.

Linaria vulgaris Miller
Common Toadflax

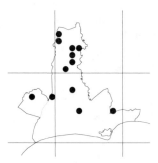

Infrequent; hedgebanks, grassland and waste places. Several scattered records, but most frequent in and around Avon Forest Park. Described by Linton and Townsend as common, especially in the Stour valley. Certainly not noticeably common here now.

Chaenorhinum minus (L.)Lange
Small Toadflax

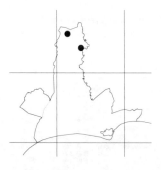

Uncommon; sandy fields and waste places. Two records, both from disturbed areas on sandy ground. Linton and Townsend both found the species uncommon in this area.

*Cymbalaria muralis Gaertn.Mey.& Scherb.
Ivy-leaved Toadflax

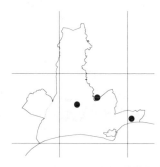

Introduced, uncommon; on old walls, usually near dwellings. Recorded on old walls near Holdenhurst village, also south of Sopley, and at Highcliffe Castle. First recorded in Britain in 1640, now common throughout most of Britain; surprisingly infrequent in this area. Recorded previously at Wick and Christchurch (Townsend).

Scrophularia nodosa L.
Common Figwort

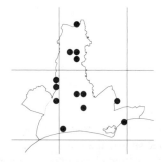

Infrequent; hedgebanks, ditches and damp woods. A few widely scattered records, nowhere common. Recorded by Linton and Townsend as common, possibly under-recorded now but certainly not now as common as then.

Scrophularia auriculata L.
Water Figwort

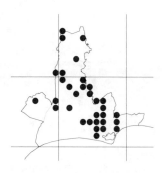

Frequent; ditches, streambanks, wet meadows and woods. Common in the Avon valley and recorded often in the Moors River and Stour valleys, and by the River Mude.

*Mimulus guttatus DC.
Monkeyflower

Introduced, uncommon; banks of rivers and streams. One site on Priory Marsh, Stanpit, on the banks of the River Avon. First recorded in Britain in 1830, it is now rather common throughout the British Isles. Previously recorded from sites in the Avon valley (ABF), but not recorded by Linton, and excluded by Townsend. 1 grid square;19

Limosella aquatica L.
Mudwort

Very rare nationally and decreasing. Found on the muddy edges of pools with a fluctuating water level. One site on Stanpit Marsh, with individual plants scattered over several metres, but very dependent upon the weather and other variables eg. grazing and water levels. Linton and Townsend recorded it as extinct in several sites in Hants. No sites recorded in the rest of Dorset.
1 grid square;19

Digitalis purpurea L.
Foxglove

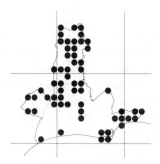

Fairly common; woods, heaths and dry banks. Recorded in most woodlands, and on sandy heaths throughout the area.

Veronica beccabunga L.
Brooklime

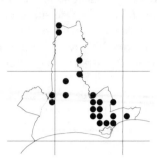

Frequent; ditches, streams and wet places. Most records from the Avon valley and around Christchurch Harbour. Perhaps not as common as described by Linton and Townsend.

Veronica anagallis-aquatica L.
Blue Water-speedwell

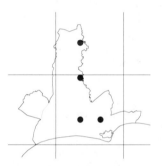

Infrequent; wet meadows and riverbanks. Several sites on the banks of the River Avon, and one record from the River Stour. Recorded by Linton and Townsend from several places by the Stour, but now it is much less common there.

Veronica catenata Pennell
Pink Water-speedwell

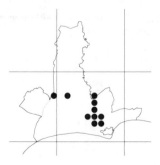

Infrequent; rivers, streams and wet meadows. Recorded only from the Avon and Stour valleys. The previous distribution of this species is not known in detail.

Veronica x lackschewitzii Kelier
(V.anagallis-aquatica x catenata)

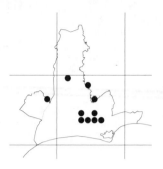

Infrequent; riverbanks and wet places, where both parents occur together. Recorded from the Avon and Stour valleys. Often more common, and usually more vigorous in growth, than the parent species.

Veronica scutellata L.
Marsh Speedwell

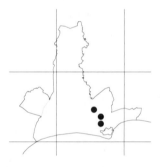

Rare; wet meadows and marshes. Three sites in the Avon valley with a few scattered plants in each one, not common anywhere. The site in Purewell Meadows may still be present, but is at risk, being now very close to roads and houses. A declining species, described by Linton as locally frequent.

Veronica officinalis L.
Heath Speedwell

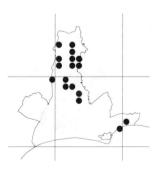

Infrequent; heaths and grassy places, especially on dry soils. Recorded on the sandy heathlands, especially in

the north of the area. Recorded by Linton and Townsend as common around Bournemouth, and in the Stour valley; somewhat decreased since then.

Veronica montana L.
Wood Speedwell

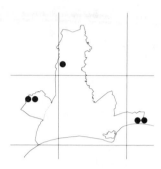

Infrequent; damp woods and shady places. All sites in damp deciduous woodland, mostly near streams. Previously recorded in old Floras from only one site in this area, at Chewton Bunny.

Veronica chamaedrys L.
Germander Speedwell

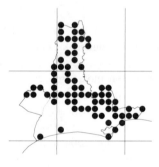

Common; hedgerows, woods and grassland. Widespread throughout the area.

Veronica serpyllifolia L.
Thyme-leaved Speedwell

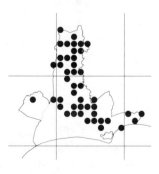

Fairly common; grassland, heaths and cultivated ground, usually in damp places. Recorded from the Stour and Avon valleys, and from sandy grassland areas.

Veronica arvensis L.
Wall Speedwell

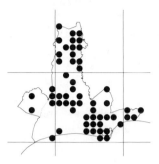

Fairly common; heaths, grassland and cultivated ground, especially on dry sandy soils. Recorded from the Stour and Avon valleys, and on sandy grasslands throughout the area.

Veronica hederifolia L.
Ivy-leaved Speedwell

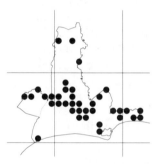

Frequent; hedges, roadsides, shady places and cultivated ground. Most frequent in the Stour and Moors River valleys, and in the woods to the east of Christchurch. Although this species was not generally determined to subspecies level the following two subsp. were recorded: *subsp. lucorum*; Broadway Lane, Throop, 1986. *subsp. hederifolia*; Holdenhurst village, 1986.

Veronica persica Poiret
Common Field-speedwell

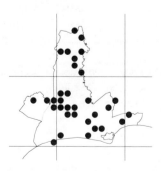

Frequent; cultivated and waste ground. Common in the Stour valley, and scattered records elsewhere throughout the area.

Veronica polita Fr.
Grey Field-speedwell
Infrequent; cultivated ground. Generally common, especially in gardens. Rather overlooked, although there are few areas of arable land locally, and gardens were excluded. One specimen found on a roadside verge near Winkton in 1987.
1 grid square;19

Veronica agrestis L.
Green Field-speedwell

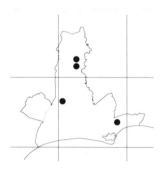

Infrequent; cultivated ground and gardens. Several sites, some on disturbed ground. Appears to be still decreasing since Linton's time.

*Veronica filiformis Sm.
Slender Speedwell

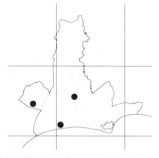

Introduced, uncommon; grassy places, roadsides and lawns. Recorded from Holdenhurst village, Turbary Common and from the grassy cliff top at Toft zigzag, Bournemouth. Not previously recorded in local Floras for this area. First recorded as an escape in England in 1927, and recorded in Dorset in 1948. Now naturalized and spreading.

Pedicularis palustris L.
Marsh Lousewort

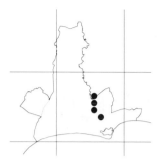

Infrequent; wet meadows and marshes. All records are from meadows in the Avon valley. Previously recorded from the Stour valley and many other local meadows, much declined now especially outside the Avon valley.

Pedicularis sylvatica L.
Lousewort

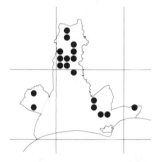

Infrequent; damp heaths, bogs and marshes. A few sites in and near the Avon valley, but most records from the bogs and heathlands in the north of the area.

Rhinanthus minor L. agg.
Yellow-rattle

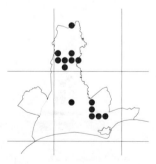

Infrequent; grassland. Recorded from several sites in and near the Avon valley, and also in and around Avon Forest Park. Described by Linton and Townsend as common, but not particularly common now.

Melampyrum pratense L.
Common Cow-wheat

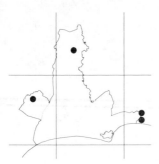

Infrequent; damp woods and thickets. All four sites in damp deciduous woodland. Fewer sites now than listed in old Floras, probably decreasing.

Euphrasia officinalis sens.lat.
Eyebright

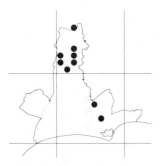

Infrequent; heaths and dry grassland. Mainly recorded on heaths and sandy grassland in the north of the area. Probably under-recorded. Described as common by Linton. *Euphrasia nemorosa* (Pers.)Wallr. was recorded in Avon Forest Park in August 1985.

Odontites verna (Bell.)Dumort.
Red Bartsia

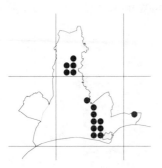

Infrequent; meadows, cultivated ground and waste places. Common in meadows in the Avon valley, and on dry grassland in and around Avon Forest Park.

Parentucellia viscosa (L.)Caruel
Yellow Bartsia

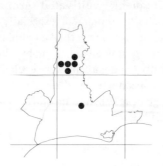

Infrequent; sandy grassland and heaths. Several good sized colonies with scattered plants around as well. Dorset and Hants are at the eastern limits of this coastal plant, common only in southwest England although increasing in its distribution in Dorset. Recorded previously in this area only at Highcliff in 1889 and near Christchurch (Townsend).

OROBANCHACEAE

Orobanche minor Sm.
Common Broomrape

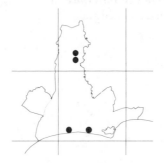

Infrequent; in dry grassland, parasitic on *Trifolium* and other leguminous plants. Two sites in Avon Forest Park on dry grassland under bracken. Only a few spikes seen at each site, in 1985 and 1989. Often eaten very quickly by rabbits. Also recorded on the cliffs at Southbourne in 1992, and Boscombe in 1993. Previously recorded in the Stour valley, and abundantly between Wick and Hengistbury in 1896 (Linton). Previously more common, and often in clover and other crops.

LENTIBULARIACEAE

Pinguicula lusitanica L.
Pale Butterwort
Uncommon; bogs and very wet heaths. Two colonies amongst *Sphagnum* moss. One site with only a few plants, the other with more, both seen in 1993. Previously recorded by Linton and Townsend as frequent, but much decreased now.
2 grid squares;09

Utricularia vulgaris agg.
Greater Bladderwort
Uncommon; ponds and ditches. Recorded in flower in a deep, shaded acid pond near St Leonards, in 1982 and 1985. Linton and Townsend recorded *U. vulgaris* and *U. neglecta* from many ditches above Christchurch, although there was some confusion as to which species was present. Also recorded from a pool on the dunes at Mudeford (Linton, 1925). Few recorded sites in Dorset.
1 grid square;10

ACANTHACEAE

***Acanthus mollis** L.
Bear's-breech
Introduced, uncommon; naturalized in waste places. Plants growing on Bournemouth East and West Cliffs amongst other vegetation, probably planted.
1 grid square;09

LABIATAE

Mentha arvensis L.
Corn Mint
Uncommon; arable fields and damp places. Two records from Avon Forest Park, on sandy grassy heathland, in 1985. Recorded as common by Linton and Townsend, but not often recorded in Dorset. Probably declined, although may be under-recorded.
2 grid squares;10

Mentha aquatica L.
Water Mint

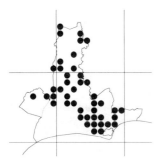

Fairly common; streams, ditches, marshes and wet places. Recorded from the Avon and Stour valleys, the Moors River and the River Mude.

Mentha x piperata var citrata (Ehrh.)Boivin
Lemon Mint
Uncommon; but widely cultivated and sometimes naturalized. A colony on a wide roadside verge near Waterditch in 1987. Probably this species, but not confirmed; most likely introduced.
1 grid square;19

***Mentha spicata** L.
Spear Mint

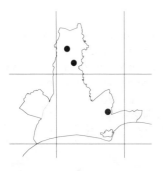

Introduced, uncommon; damp roadsides and waste places. Two records from damp roadside verges in 1987 and one from damp heathy grassland in 1985. Not recorded in old Floras for this area, and few records in Dorset, although it could occur anywhere as an escape from cultivation.

Mentha x villosa Hudson
Apple Mint
Uncommon; often naturalized. A single record from a roadside verge near Holdenhurst in 1981.
1 grid square;19

Mentha suaveolens Ehrh.
Round-leaved Mint
Uncommon; ditches, roadsides and waste places in southwest England and Wales. A single site with several plants by the River Mude near Christchurch, in 1987. No local sites listed in old Floras.
1 grid square;19

Lycopus europaeus L.
Gipsywort

Fairly common; riverbanks, ditches, marshes and wet meadows. Recorded mainly in the Avon, Stour and Moors River valleys and along the River Mude.

Calamintha sylvatica subsp. ascendens (Jordan)P.W.Ball
Common Calamint
Uncommon; dry banks usually calcareous. A few plants in 1986 on a dry bank in a lane near Holdenhurst. Recorded near Sopley by Linton, and in 1926 (Rayner).
1 grid square;19

Clinopodium vulgare L.
Wild Basil
Uncommon; hedges, scrub and woodland edges, usually on calcareous soils. One site near Palmersford in rough ground near a track. Not recorded for this area in old Floras, but generally common in southern England.
1 grid square;00

Salvia pratensis L.
Meadow Clary
Uncommon; sometimes naturalized or as a casual, very rare as a native. An atypical form of this species found in 1991, on the edge of a car park near Hurn, probably a casual from dumped garden material.
1 grid square;19

Salvia verbenaca L.
Wild Clary
Uncommon; dry banks and grassland especially near the coast. A small colony recorded in 1993 on the grassy mound of the keep at Christchurch Castle, where it was recorded by Wise before 1863 (Townsend). Also recorded by Linton and Townsend as frequent about Christchurch, around Bournemouth and at Purewell.
1 grid square;19

Prunella vulgaris L.
Selfheal

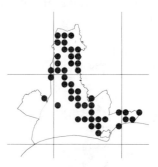

Fairly common; grassland, hedges, woods and waste places. Recorded widely throughout the area, in the Avon valley, and in most woods and grasslands.

Stachys arvensis (L.)L.
Field Woundwort

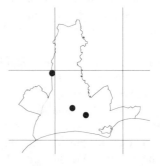

Uncommon; arable fields and waste places. One site on waste ground by a heathland track, another on waste ground near the River Stour and the third on a roadside hedgebank near Holdenhurst. Many local records listed in Linton, but much decreased in this area now.

Stachys palustris L.
Marsh Woundwort

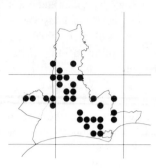

Frequent; streamsides, ditches and wet places. Mainly recorded in the Stour, Avon and Moors River valleys, but also in hedgerows and ditches elsewhere.

Stachys sylvatica L.
Hedge Woundwort

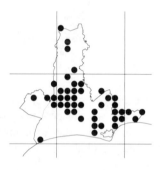

Fairly common; woods, hedges and ditchbanks. Recorded in the river valleys and in most woodlands, but much less common on the dry sandy soils especially in the north of the area.

Stachys x ambigua Sm. **(Stachys palustris x sylvatica)**
Hybrid Woundwort
Uncommon; probably often overlooked. Generally widespread nationally, with either both, one or neither parents. One site by the River Stour at Throop, in 1982. Not recorded by Linton or Townsend for this area. 1 grid square;19

Stachys officinalis (L.)Trev.
Betony
Uncommon; open woods, and hedgebanks. A single record from damp meadows at Highcliffe in 1989. Previously recorded from Bournemouth and Christchurch by Townsend, and described as common. 1 grid square;29

*Stachys lanata Jacq. non Crantz
Lamb's-ear
Introduced, uncommon; commonly grown in gardens, often a persistent throw-out or self-sown. Two separate plants recorded in 1993 on a ride in a forestry plantation at Ashley Heath. 1 grid square;10

Ballota nigra L. **subsp. foetida** Hayek
Black Horehound

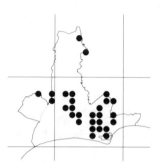

Frequent; roadsides, hedgebanks and waste places, especially near buildings. Most common in the Stour and Avon valleys, and around Bockhampton.

Lamium amplexicaule L.
Henbit Dead-nettle

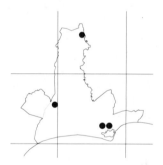

Infrequent; cultivated and waste ground. One record as a weed in Muscliff, North Bournemouth; two sites on rough ground near Purewell, Christchurch, and one record in a field near Ashley. Previously recorded as common in the Stour valley, and also at Wick, Christchurch and Mudeford in 1879 (Townsend, Linton). Not common now.

Lamium hybridum Vill.
Cut-leaved Dead-nettle

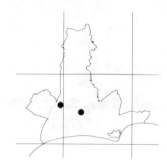

Rare; in cultivated ground. Two sites, both near Bournemouth; one on a dry bank near Holdenhurst in 1986; the other in a flower bed at the front of Epiphany school, Muscliff in 1991. Few recent records in Dorset.

Lamium purpureum L.
Red Dead-nettle

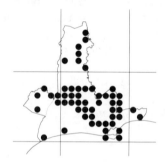

Fairly common; hedges, waste places and cultivated ground. Common in the Stour valley and in hedgerows to the east of the River Avon.

Lamium album L.
White Dead-nettle

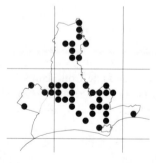

Fairly common; hedges, roadsides, disturbed ground and waste places. Recorded in the Stour and Avon valleys, and in hedgerows to the east of the River Avon; also around Hurn and in Avon Forest Park.

Galeopsis tetrahit agg.
Common Hemp-nettle

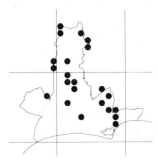

Scutellaria galericulata L.
Skullcap

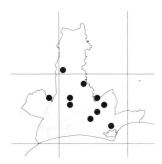

Teucrium scorodonia L.
Wood Sage

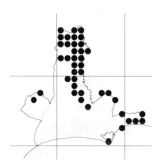

Frequent; cultivated and waste ground, woods, hedges and ditches. Records scattered throughout the area, many in ditches and damp grassy places. *G. tetrahit subsp. tetrahit* recorded in 1987 by the River Mude.

Infrequent; streambanks, ditches and wet meadows. Recorded in the Stour, Moors River and Avon valleys. Described by Linton and Townsend as locally common, it seems to be less common now.

Fairly common; heaths, woods and hedges on drier sandy soils. Common on the drier sandy soils in the north of the area, and recorded in the east along the coast.

Galeopsis bifida Boenn.
Lesser Hemp-nettle
Uncommon; arable land and cultivated fields. Two records, one near Mudeford Wood and one near the Moors River north of Hurn Airport. Other records have been included in *G. tetrahit* agg.
2 grid squares;19

Scutellaria minor Hudson
Lesser Skullcap

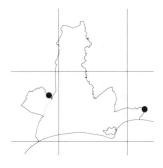

Ajuga reptans L.
Bugle

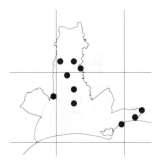

Glechoma hederacea L.
Ground-ivy

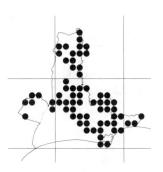

Uncommon; wet heaths and damp sandy places. A record in 1986 on a damp heath which had been burnt a few years before; and also one record nearby, just outside the area (in Hants), also from a damp heath. Also recorded in Ensbury Wood, Bournemouth in 1993. Described by Linton and Townsend as frequent and rather common, but much decreased since then.

Infrequent; damp woods and meadows. A few records from wet grasslands in the Stour, Avon and Moors River valleys, and one or two from damp deciduous woodlands. Recorded by Linton and Townsend as frequent and common, but certainly not common now.

Common; woods, hedgerows and grassland. Most records in the Stour valley and to the east of the River Avon, less common on the drier soils.

PLANTAGINACEAE

Plantago major L.
Great Plantain

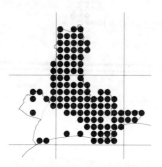

Very common everywhere; grassland, roadsides and disturbed ground. Recorded in almost all grid squares.

Plantago lanceolata L.
Ribwort Plantain

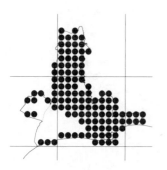

Very common everywhere; grassy places and disturbed ground. Recorded in all but four grid squares, almost certainly present in these also.

Plantago maritima L.
Sea Plantain

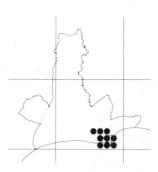

Infrequent; saltmarshes, mudflats and estuaries. Recorded only in Christchurch Harbour and around Hengistbury Head.

Plantago coronopus L.
Buck's-horn Plantain

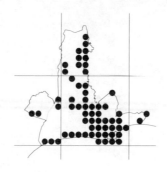

Common; heaths, grassland and sandy and gravelly places. Very common along the coast and around Christchurch Harbour. Also recorded in the Stour and Avon valleys and on heaths, usually on sandy and gravelly tracks.

Littorella uniflora (L.)Ascherson
Shoreweed
Rare; margins of pools on sandy and heathy ground. A single colony seen in August 1991 with several dozen plants on the muddy edge of a dried up pond. Few sites in Hants (Townsend) or Dorset. Previously recorded at Herne (Linton).
1 grid square;19

CAMPANULACEAE

Campanula trachelium L.
Nettle-leaved Bellflower
Uncommon; usually on chalk. A single 'casual' plant in 1986 growing on the edge of a car park near Hurn, probably a garden 'throwout'. Not recorded by Linton or Townsend for this area.
1 grid square;19

Campanula persicifolia L.
Peach-leaved Bellflower
Introduced, uncommon; a casual in open woods and commons, extinct as a native. Several plants growing well in 1986 on the edge of a car park at Hurn, possibly a garden 'throwout'. Also a single plant on a wide trunk road verge flourishing in 1991 - probably a garden escape.
2 grid squares;10,19

Campanula rotundifolia L.
Harebell
Infrequent; dry grassy places and sandy heaths. Three sites on dry sandy heathy grassland, and one site on dry disturbed grassland, with few plants at each site, therefore all vulnerable. Described as common in Linton and Townsend, so obviously decreased considerably since then.
4 grid squares;19

Jasione montana L.
Sheep's-bit

Infrequent; sandy heathlands and grassy places. Recorded in and around Avon Forest Park, and along the coast near Hengistbury Head. Previously frequent, probably decreased now. Only common in Dorset in the Poole Basin.

RUBIACEAE

Sherardia arvensis L.
Field Madder

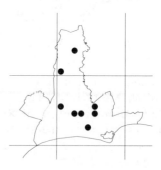

Infrequent; roadsides, cultivated and waste ground. Most records in the Stour and Avon valleys in sandy and gravelly places and on dry grassland. Recorded by Linton and Townsend as common, certainly not common now.

Galium mollugo L. sens.lat.
Hedge-bedstraw

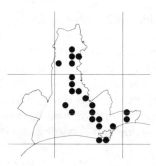

Frequent; hedges, woods and grassy places, especially on calcareous soils. Recorded in the Avon and Moors River valleys, and in Avon Forest Park.

Galium verum L.
Lady's Bedstraw

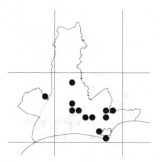

Infrequent; hedges, dry banks and grassy places especially on calcareous soil. Most records from the Stour valley, and north of Somerford. Not as frequent as described in Linton and Townsend.

Galium saxatile L.
Heath Bedstraw

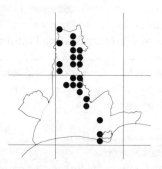

Frequent; heaths and dry grassy places on acid soils. Recorded mainly from the dry sandy heaths in the north of the area and around Hurn.

Galium palustre L. sens.lat.
Common Marsh-bedstraw

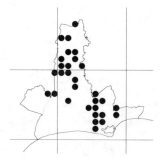

Frequent; ditches, streambanks, marshes and wet places. Most frequent in the Avon valley, also recorded in the Stour and Moors River valleys. Recorded by Linton and Townsend as very common, perhaps decreased somewhat.

Galium uliginosum L.
Fen Bedstraw

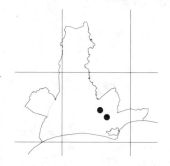

Uncommon; wet meadows and marshes. Two sites, both in marshy meadows in the Avon floodplain. Previously common in both the Avon and Stour valleys; probably decreasing, due to drainage of habitats and more intensive agricultural use.

Galium aparine L.
Goosegrass, Cleavers

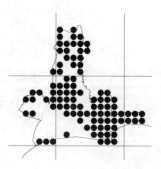

Very common; hedges, thickets, cultivated and waste ground. Widespread throughout the area, but less common in the drier areas.

CAPRIFOLIACEAE

Sambucus nigra L.
Elder

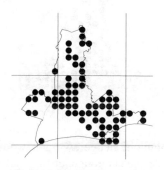

Common; woods, roadsides and waste places. Recorded in woods and hedgerows in most areas, but less common on the drier sandy soils in the north of the area.

Viburnum opulus L.
Guelder-rose

Infrequent; woods, copses and hedgerows. A few records in the Stour valley and around Hurn. Described as frequent in Linton and Townsend for this area, but not as frequent now.

Symphoricarpos albus (L.) S.F.Blake
Snowberry

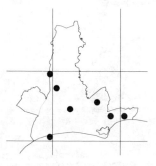

Introduced, infrequent; often planted and widely naturalized. A few scattered records, mainly in woods, thickets and hedgerows.

Lonicera periclymenum L.
Honeysuckle

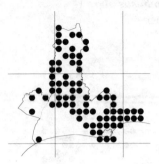

Common; hedgerows and woods. Widespread throughout the area, but more common to the east of Christchurch.

ADOXACEAE

Adoxa moschatellina L.
Moschatel
Uncommon; woods and shady hedgebanks. Three sites with good sized colonies, all in deciduous woodland by streams or rivers. Recorded by Linton and Townsend in only two sites in the area.
3 grid squares;09,19

VALERIANACEAE

Valerianella locusta (L.)Laterrade
Common Cornsalad
Uncommon; arable land, hedgebanks and disturbed ground. One record in 1987 from roadside verges by arable fields at Holfleet, Winkton. Possibly under-recorded but certainly not frequent now, as recorded by Linton.
1 grid square;19

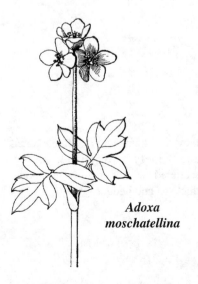

Adoxa
moschatellina

Valerianella carinata Loisel.
Keeled-fruited Cornsalad

Uncommon; arable fields, banks and old walls. One record in 1990 from East Cliff, Bournemouth and one at Ashley in 1993. Not recorded previously in this area in local Floras, few records in Dorset, but possibly under-recorded.
1 grid square;09

Valeriana officinalis L.
Common Valerian

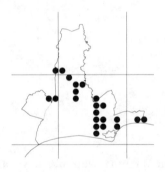

Frequent; wet meadows and marshy places. Recorded mainly in the Avon valley, in the Moors River valley and by the River Mude. Not now common in the Stour valley, as recorded by Linton.

Valeriana dioica L.
Marsh Valerian

Uncommon; wet and marshy meadows. Two sites, both in very wet marshy meadows by the River Avon. Recorded by Linton and Townsend as common in meadows by the Stour and its tributaries. Much declined

now, and apparently restricted to the Avon valley.

*_Centranthus ruber_ (L.)DC.
Red Valerian
Introduced, infrequent; cliffs, old walls and rocks, especially near the sea. Surprisingly only one record, at Iford, of this frequently cultivated species. Normally well naturalized on walls, cliffs and dry banks in south and west England; however Townsend and Linton did not record this species around Bournemouth.
1 grid square;19

DIPSACAEAE

Dipsacus fullonum L.
Teasel

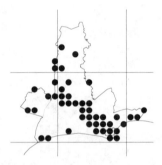

Fairly common; damp hedgerows, streambanks and roadsides. Common in the Stour valley, with a few records elsewhere. Previous records also mainly in the Stour valley (Linton).

Knautia arvensis (L.)Coulter
Field Scabious
Uncommon; dry rough grassland on chalk and sand. A single record from Avon Forest Park South in 1986 in dry grassland. Described in Linton and Townsend as common in field borders and banks. Much decreased now.
1 grid square;10

Scabiosa columbaria L.
Small Scabious
Uncommon; usually in dry calcareous pastures and banks. One site, on a dry bank in Avon Forest Park; a surprising place for this calcicole. However a high lime content can result from broken seashells, roadworks, old buildings etc. which, on a dry sandy soil would provide similar conditions to a chalk soil. Not recorded in this area by Linton or Townsend.
1 grid square;10

Succisa pratensis Moench
Devil's-bit Scabious

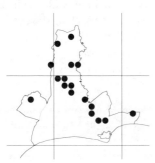

Infrequent; damp meadows and heaths. Recorded mainly in wet meadows in the Avon and Moors River valleys, but also on heathland at Kinson, Highcliffe and Avon Forest Park. Recorded by Linton and Townsend also in the Stour valley.

COMPOSITAE

*Helianthus annuus L.
Sunflower
Introduced, uncommon; usually an escape from cultivation. Two records, one on river banks at Iford near a caravan park and golf course, and one at Hengistbury Head near beach huts. 2 grid squares,19

Bidens cernua L.
Nodding Bur-marigold

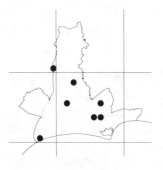

Infrequent; pond and streamsides and wet places. A few records from the Avon, Stour and Moors River valley. Recorded by Linton and Townsend as common; Townsend considered it to be more common than *B. tripartita*.

Bidens tripartita L.
Trifid Bur-marigold

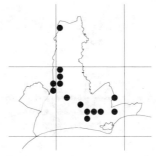

Infrequent; ponds, ditches, streamsides and marshy places. Recorded mainly in the Stour valley, and by the River Mude. More common than *B. cernua*. The distribution in Dorset is described by Good as being chiefly in the River Stour system.

*Galinsoga parviflora Cav.
Gallant Soldier

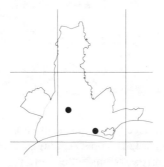

Introduced, uncommon; arable land and waste places. One site in North Bournemouth on waste ground, the other near Wick as a weed in cultivated ground. Well established around London, but still not common in this area, although probably spreading. Scarce in Dorset, only two sites mentioned in Good (1948). Not recorded in Townsend or Linton.

*Galinsoga ciliata (Rafin.)S.F.Blake
Shaggy Soldier

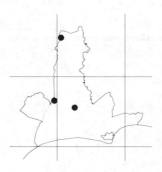

Introduced, infrequent; cultivated ground and waste places. All records are from cultivated ground, near

habitation. First recorded in Britain in 1919, and apparently recorded in the Christchurch area before 1940; this species is now spreading fast as a weed generally, but not apparently, locally. Recorded in 1962 at Bournemouth Bus Station (PBNSS.52,57).

Senecio jacobaea L.
Common Ragwort

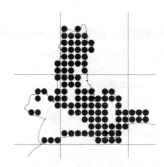

Very common everywhere; meadows, roadsides and waste and disturbed ground. Not recorded from a few grid squares, particularly in the mainly arable areas east of Burton.

Senecio aquaticus Hill
Marsh Ragwort

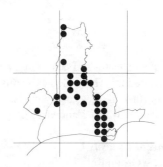

Frequent; wet meadows and marshes. Most common in the Avon valley, but also recorded from the Stour and Moors River valleys.

*Senecio squalidus L.
Oxford Ragwort

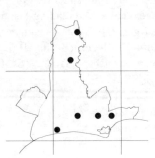

Introduced, infrequent; walls, waste places and disturbed ground. A few widely scattered records. Although first recorded in Oxford in 1794, not mentioned by Townsend or Linton. Some Dorset records around Poole (Good, 1984).

Senecio sylvaticus L.
Heath Groundsel

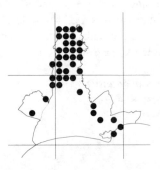

Frequent; heaths and sandy grassland. Recorded mainly on the sandy soils in the north of the area, and some records on dry sandy grassland elsewhere.

Senecio viscosus L.
Sticky Groundsel

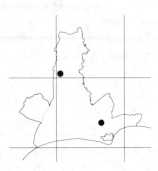

Uncommon; waste ground, railway banks and tracks. One record from the railway bank east of Christchurch, and one from disturbed sandy heathland areas. Not recorded by Linton or Townsend in this area, some records for the southeast of Dorset (Good, 1984).

Senecio vulgaris L.
Groundsel

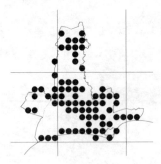

Common; cultivated and waste ground. Widespread throughout the area, although more common in the cultivated areas and along the coast.

*Senecio cineraria DC. sens.lat.
Silver Ragwort

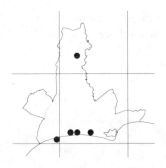

Introduced, infrequent; often grown in gardens and escaping. Several records from the cliffs at Bournemouth; and one record in Avon Forest Park, almost certainly a garden outcast.

*Doronicum pardalianches L.
Leopard's Bane
Introduced, uncommon and usually planted; woods and copses. A single 'casual' plant flowering in 1989 in a hedgerow in Walkford. Probably from dumped garden material, growing with *Geranium endressii* and *Meconopsis cambrica*.
1 grid square;29

Tussilago farfara L.
Colt's-foot

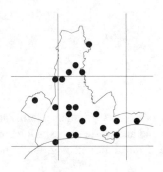

Frequent; banks and waste ground especially on heavy soils. Recorded mainly in the Stour valley, but also in several places along the coast.

*Petasites fragrans (Vill)C.Presl.
Winter Heliotrope

Introduced, infrequent; banks, streamsides and waste places. One site on the old rubbish tip by Priory Marsh, Christchurch in 1986, some on the East Cliff, Bournemouth in 1987, and cliff tops by Middle Chine, 1991. Also at Highcliffe Castle. Not recorded locally by Linton or Townsend, but now becoming more frequent in Dorset and Hants.

*Calendula officinalis L.
Marigold

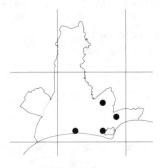

Introduced; casuals or garden escapes. Several records as a garden escape, and on the cliffs at Boscombe. One record from Stanpit Marsh in 1986, where it was growing by the river bank.

Pulicaria dysenterica (L.)Bernh.
Common Fleabane

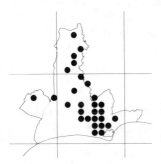

Frequent; marshes, wet meadows and damp places. Most common in the Avon valley but also recorded frequently in the Stour valley.

Pulicaria vulgaris Gaertner
Small Fleabane
Very rare and decreasing nationally; moist sandy places and pond margins. A single colony on a farm track in the Avon valley, seen in 1993. No other Dorset sites, a few in the New Forest. Previously recorded by Linton and Townsend from sites around Bournemouth, and near Christchurch in 1835.
1 grid square

Filago vulgaris Lam.
Common Cudweed

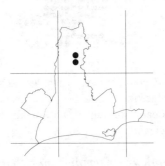

Uncommon; dry heaths and pastures on acid sandy soils. Two records, both from Avon Forest Park in 1985, on dry heathy grassland. Described by Linton and Townsend as common in dry cultivated fields; however, although these habitats have decreased locally in recent years, even in the most likely habitats this species is now uncommon.

Filago minima (Sm.)Pers.
Small Cudweed

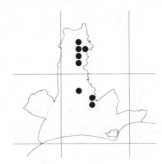

Infrequent; dry sandy grassland and heaths. Recorded from Town Common and Avon Forest Park. Described as common in Linton, possibly under-recorded now.

Gnaphalium uliginosum L.
Marsh Cudweed

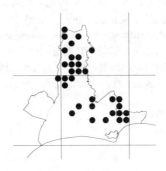

Frequent; damp places in sandy fields and heaths. Recorded mainly from the sandy heaths and grasslands in the north of the area, and in the arable fields to the east of Burton.

Solidago virgaurea L.
Goldenrod

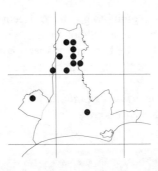

Infrequent; sandy heaths and dry grassland. Some records from Avon Forest Park and surrounding dry sandy heaths and grasslands, very few records elsewhere. Recorded by Linton and Townsend as common, apparently decreased since then.

***Solidago canadensis** L.
Garden Goldenrod

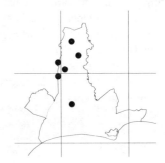

Introduced, infrequent; often cultivated and sometimes escaping. A few scattered records, mainly on the drier sandy soils.

Aster tripolium L.
Sea Aster

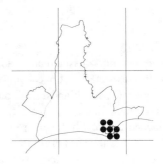

Infrequent; saltmarshes and seashores. Common on Stanpit Marsh, and recorded around Christchurch Harbour.

***Aster novi-belgii** L. sens.lat
Michaelmas-daisy

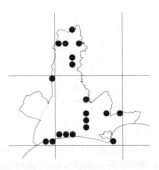

Introduced, infrequent; much grown in gardens and often an escape. Several scattered records, and many plants growing on the cliffs from Durley Chine to Boscombe.

*Aster lanceolatus Willd.
Narrow-leaved Michaelmas-daisy
Introduced, uncommon; a garden escape. Several plants, probably this species or possibly A. *x salignus*, recorded on waste ground in North Bournemouth. This species has apparently been grown in gardens since the 17th century.
1 grid square;19

Erigeron acer L.
Blue Fleabane

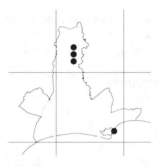

Infrequent; dry grassland and sandy places, especially on calcareous substrata. On heathlands, but only found where there is a calcareous influence, noticeably absent from the others. Described as not common in Linton and Townsend, and only listed for one site in the area, around Highcliff.

*Erigeron glaucus Ker-Gawker
Seaside Daisy

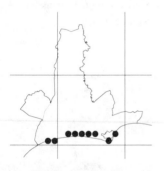

Introduced, infrequent; a garden escape and sometimes established. One site with a reasonable colony on the sand dunes at Hengistbury Head, and several records on the cliffs between Durley Chine and Southbourne. Not mentioned in local Floras until Good, 1984.

*Conyza canadensis (L.)Cronq.
Canadian Fleabane

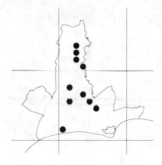

Introduced, infrequent; waste and cultivated ground on sandy soils. Recorded in and around Avon Forest Park, on Town Common, and on cliffs near Boscombe. Not recorded in this area by Townsend or Linton. Townsend listed it as rare, in only two Hants. sites, and stated "This is a plant which spreads very rapidly when once introduced, and it will in all probability soon become abundant". Rayner listed it at Christchurch in 1924 and 1926. First recorded in Dorset in 1917, now locally common.

Bellis perennis L.
Daisy

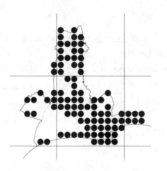

Very common; grassland. Recorded in most areas, although it seems to prefer the heavier soils, and is less common in the north of the area.

Eupatorium cannabinum L.
Hemp-agrimony

Frequent; streambanks, ditches and wet places. Recorded in the Stour,

Avon and Moors River valleys, by the River Mude and in ditches elsewhere.

Achillea ptarmica L.
Sneezewort

Infrequent; damp meadows and heathy grassland. Several records in damp meadows in the Moors River valley and from sandy grassland nearby. Recorded as frequent and common by Linton and Townsend in the Stour watershed.

Achillea millefolium L.
Yarrow

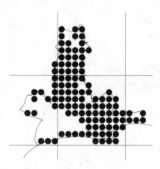

Very common everywhere; grassland, hedgerows and disturbed ground. Probably present in almost all grid squares.

Tripleurospermum maritimum sens.lat.
Mayweed

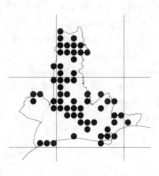

Common; cultivated and waste ground, and coasts. Widespread throughout the area, and along the coast.

Tripleurospermum maritimum
(L.)Koch sens.str.
Sea Mayweed
Uncommon; dunes, shingle beaches and seashores. Recorded from the shingle beach at Hengistbury Head in 1991. Probably common along the whole length of coast. Earlier records are included in *T. maritimum* sens.lat.
1 grid square;19

Tripleurospermum inodorum
(L.)Schultz Bip.
Scentless Mayweed
Generally common; cultivated and waste ground. Recorded at Purewell Cross by the car park, and on Spellars Point, Stanpit Marsh in 1991. Also recorded near Ashley. Most records are included in *T. maritimum* sens.lat.
2 grid squares;19

Matricaria recutita L.
Scented Mayweed

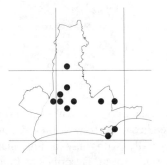

Infrequent; sandy arable and waste ground. A few scattered records, and several in the Stour valley.

*Matricaria matricarioides
(Less.)Porter
Pineappleweed

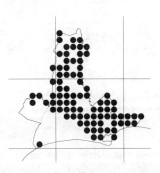

Introduced, very common; waysides, tracks and waste places. Not mentioned in Linton or Townsend. First recorded in Hants near the Mill at Christchurch in 1903 (Linton, 1919). Rayner describes it in 1929 as "now seen everywhere on roadsides and waste places"

Chrysanthemum segetum L.
Corn Marigold

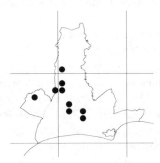

Infrequent; sandy arable fields. A few records from the Stour valley and nearby. Described as common by Linton, but much decreased since then.

Leucanthemum vulgare Lam.
Oxeye Daisy

Frequent; grassland, banks and roadsides. Many scattered records but probably more frequent in the north of the area. Described by Linton and Townsend as common and abundant.

*Tanacetum parthenium
(L.)Schultz Bip.
Feverfew

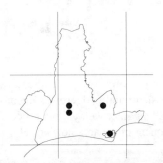

Introduced, infrequent; waste places and near buildings. Several sites, all on waste ground or very close to buildings. Formerly cultivated as a medicinal herb and used against fevers.

Tanacetum vulgare L.
Tansy

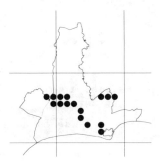

Infrequent; riverbanks, hedges and roadsides. Recorded often in the Stour valley, and a few records near Bockhampton. Recorded by Linton and Townsend in the Stour valley and at Pokesdown.

Artemisia vulgaris L.
Mugwort

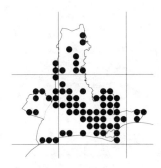

Common; hedges, roadsides and waste places. Recorded mainly in the Stour valley, east of Burton and along the coast.

Arctium lappa L.
Greater Burdock

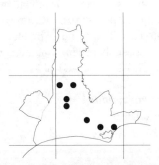

Infrequent; tracks, rough and disturbed ground, often near rivers. Most sites by the River Stour or its tributaries, a few on disturbed ground nearby. Not recorded in this area by Linton and Townsend, described as rare.

Arctium minus agg.
Lesser Burdock

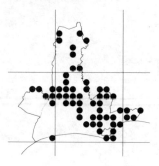

Fairly common; hedgerows, riverbanks, copses and waste ground. Recorded mainly in the Stour and Avon valleys and in hedgerows and woods mainly to the east of the River Avon.

Carduus tenuiflorus Curtis
Slender Thistle

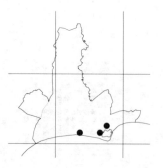

Uncommon; in disturbed ground and waste places near the sea. Two sites on Stanpit Marsh with several strong plants. Many plants growing on the shingle 'beach' by Grimmery Bank in 1986. Also recorded in 1993 on the cliffs at Boscombe. Surprisingly few sites as it is common around most of England near the sea. Recorded by Linton and Townsend as frequent on the coast between Bournemouth and Christchurch and at Mudeford.

Carduus nutans L.
Musk Thistle

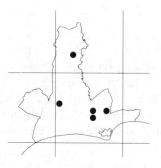

Infrequent; grassland, hedges and waste places, usually on calcareous

soils. A few widely scattered records, not plentiful anywhere. Not recorded by Linton for this area.

Cirsium eriophorum (L.)Scop.
Woolly Thistle
Uncommon; grassland, open waste ground and roadsides on calcareous soils. One record in 1985 of a single very large plant in Avon Forest Park South, growing with other species favouring calcareous soils. No sites mentioned locally in old Floras.
1 grid square;10

Cirsium vulgare (Savi)Ten.
Spear Thistle

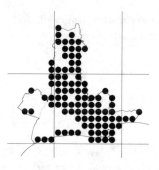

Very common; grassland, roadsides and disturbed ground. Widespread throughout the area.

Cirsium palustre (L.)Scop.
Marsh Thistle

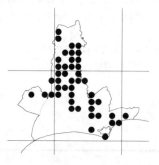

Fairly common; marshes, wet grassland and hedgerows. Recorded in all the river valleys, and especially frequent in the area north of Hurn.

Cirsium arvense (L.)Scop.
Creeping Thistle

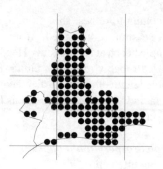

Very common everywhere; grassland, waste places and cultivated ground. Present in almost all grid squares, though less frequent in the northern part of the area.

Cirsium dissectum (L.)Hill
Meadow Thistle

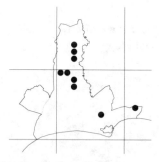

Infrequent; damp heaths and boggy meadows. A few scattered sites, generally with only a few plants present at each. Described by Linton and Townsend as rather common, certainly much decreased now.

Centaurea scabiosa L.
Greater Knapweed

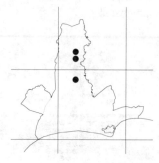

Infrequent; dry grassland and roadsides. Common on calcareous soils in southern England. Locally found growing on dry heathy grassland, with other species also often associated with calcareous soils. Recorded by Townsend and Linton at Bournemouth, near Christchurch and at Highcliff.

Centaurea cyanus L.
Cornflower
Rare; an occasional casual of waste places and cornfields. A single 'casual' plant seen in 1986, in Mill Road, Bournemouth. Perhaps an escape from cultivation, although the site is less than 1 mile from Muscliff where it was recorded by Linton and Townsend. They also noted it in the Stour watershed. Much rarer now, due to more intensive agriculture, including the better cleaning and sorting of arable seeds, and the use of herbicides. Nationally declined dramatically from 264 to 3 10km grid squares (excluding casual records) between 1930 and 1990 (Wilson, 1992).
1 grid square;19

Centaurea nigra L. agg.
Common Knapweed

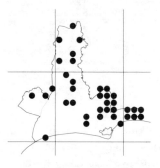

Frequent; hedgerows, roadsides and grassland. Most common in the hedgerows to the east of Burton and around Highcliffe, with scattered records in the Stour and Moors River valleys and elsewhere.

Centaurea nigra subsp. nemoralis (Jordan)Gugler
Slender Knapweed

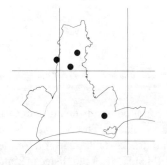

Infrequent; usually on light calcareous soils. All four sites on heathy grassland, growing with other plants favouring chalky soils. Probably under-recorded with many plants included in the aggregate *C. nigra*. Not listed by Linton or Townsend.

Serratula tinctoria L.
Saw-wort

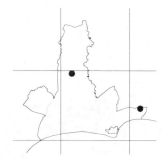

Uncommon; wood edges and damp heaths. One site recorded in 1981 in a damp meadow by a wood edge, the other with several plants growing on a previously burnt area of a damp heath in 1989. Recorded by Linton and Townsend more frequently, so probably declined since then.

Cichorium intybus L.
Chicory
Uncommon; cultivated field borders and roadsides, usually on calcareous soils. One record from North Bournemouth on a verge at the side of an unmade track. Frequently recorded in Linton for this area, much declined now.
1 grid square;19

Lapsana communis L.
Nipplewort

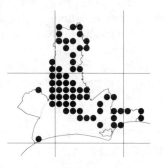

Common; hedgerows, roadsides, cultivated and waste ground. Recorded in all the river valleys and particularly common around Throop, also scattered records elsewhere.

Hypochoeris radicata L.
Cat's-ear

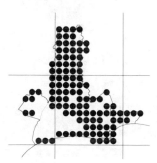

Very common everywhere; grassland, hedges, roadsides and disturbed ground. Widespread throughout the area.

Leontodon autumnalis L.
Autumn Hawkbit

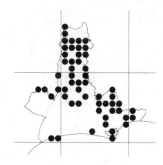

Fairly common; grassland, roadsides and waste places. Recorded frequently throughout the area, but more common on the drier sandy soils.

Leontodon hispidus L.
Rough Hawkbit

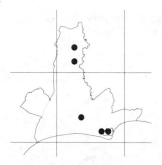

Infrequent; grassland and hedgerows, especially on calcareous soils. Two records in Avon Forest Park, one at Iford and two on Stanpit Marsh.

Leontodon taraxacoides (Vill.)Merat
Lesser Hawkbit

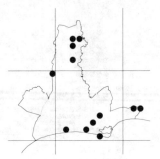

Infrequent; dry sandy grassland and heaths. Recorded in Avon Forest Park, and in several places near the coast. Described in Linton and Townsend as common, probably under-recorded now.

Picris echioides L.
Bristly Oxtongue

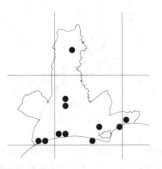

Infrequent; roadsides, hedges and disturbed ground, especially on clays and calcareous soils. Mainly recorded along the coast. Not recorded in Linton and Townsend for this area.

Tragopogon pratensis L.
Goat's-beard

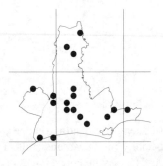

Infrequent; grassland, roadsides and waste places. Several records in the Stour valley, and also some scattered ones.

Lactuca serriola L.
Prickly Lettuce

Frequent; waste places and on walls. Recorded mainly in the Stour valley, common in some places. Not recorded by Linton or Townsend, and only three sites mentioned in Dorset in Good (1948). Much increased since then. First recorded in Dorset in 1937 and locally common now.

Sonchus arvensis L.
Perennial Sow-thistle

Infrequent; cultivated and waste ground, and on shores. Recorded around Hengistbury Head, and in a few other scattered sites. Recorded as locally common by Linton and Townsend, certainly not so common now.

Sonchus oleraceus L.
Smooth Sow-thistle

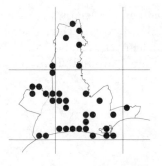

Frequent; hedgerows, cultivated and waste ground. Recorded mainly in the Stour valley and along the coast.

Sonchus asper (L.)Hill
Prickly Sow-thistle

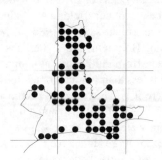

Common; hedgerows, cultivated and waste ground. Recorded in all the river valleys, along the coast and in scattered sites elsewhere.

Hieracium agg.
Hawkweed

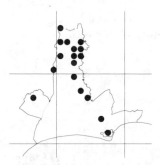

With the exception of _Hieracium pilosella_, the species of _Hieracium_ named were assigned to the most likely section or group (type), and are in no way definite. Apart from the agg. records, the genus is undoubtedly under-recorded for this area.

Hieracium umbellatum L.

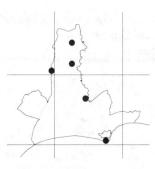

Uncommon; sandy heaths, dunes and grassy places. Three records, all from tracks and paths on or near heathland. Many sites listed in Linton, mainly on heaths and commons.

Hieracium sabaudum L.

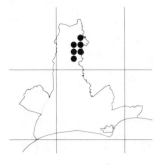

Infrequent; sandy heathland. All the sites recorded are in Avon Forest Park, other sites in the area are undoubtedly present.

Hieracium trichocaulon
(Dahlst)K.Joh.
Uncommon; sandy heaths and roadsides. Recorded from the Avon Causeway on the edge of Town Common in 1986.
1 grid square;19

Hieracium pilosella L.
Mouse-ear Hawkweed

Frequent; sandy grassland and heaths. Recorded from most of the dry sandy grasslands in the area.

*Hieracium brunneocroceum
Pugsley
Fox-and-cubs

Introduced, uncommon; roadsides, banks and waste places. Several sites, all on dry grassy heathland, with many scattered plants. Not recorded in this area by Linton or Townsend, and only recorded from six localities in Dorset by Good (1984), but recorded more frequently recently.

Crepis vesicaria L. **subsp. haenseleri**
(Bois.ex DC.)
Beaked Hawk's-beard

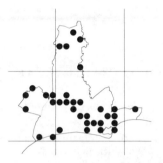

Frequent; grassland, roadsides, waste places and cultivated ground. Recorded mainly in the Stour valley, by the River Mude, and along the coast. Linton described this species as scarce, but increasing; and Rayner states that it had enormously increased in every part of the county (Hants). Very common in some areas now, and probably still increasing.

Crepis capillaris (L.)Wallr.
Smooth Hawk's-beard

Frequent; grassland, heaths and waste places. Records scattered throughout the area, especially on the drier soils.

*Crepis nicaeensis Balbis
French Hawk's-beard
Introduced; a 'casual' of arable and grass fields. One record on a verge of the new road to the Visitor Centre, Avon Forest Park in 1987. Not recorded by Linton or Townsend for this area.
1 grid square;10

Taraxacum agg.
Dandelion

Very common everywhere; grassland, roadsides and waste places. Present in almost all grid squares.

Taraxacum section **erythrosperma**
Dahlst.

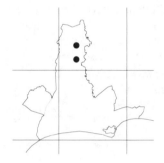

Uncommon; dry grassland and sandy heaths. Two records from Avon Forest Park in 1985. Undoubtedly many more present in other suitable locations.

ANGIOSPERMAE - MONOCOTYLEDONES

ALISMATACEAE

Alisma plantago-aquatica L.
Water-plantain

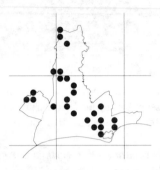

Frequent; river and streambanks, pools and ditches. Recorded in the Stour, Avon, Moors River and the River Mude.

Sagittaria sagittifolia L.
Arrowhead

Infrequent; streams and rivers, especially slow-flowing ones. Recorded in the Stour and Avon Rivers, but more frequent in the Moors River. Linton and Townsend recorded it in many places in the Stour and Avon and around Christchurch. Apparently decreased since then.

BUTOMACEAE

Butomus umbellatus L.
Flowering-rush

Infrequent; margins of rivers, and nearby ditches. All records are from the lower parts of the Rivers Avon and Stour, or in pools in meadows adjacent to them. Apparently limited to the main rivers and not recorded in the tributaries. Recorded by Linton and Townsend in the River Avon and nearby ditches, and at Wick. Very uncommon in Hampshire and Dorset outside the larger rivers.

HYDROCHARITACEAE

Hydrocharis morsus-ranae L.
Frogbit
Rare, and decreasing; ponds and ditches, usually calcareous. Several plants recorded in 1993 in the pond at the Hengistbury Head Outdoor Centre. These plants which were introduced in August 1992 originally came from a site in Christchurch Meadows which has been an industrial development for the last 20 years or more (pers.comm. J.Lavender). There are no recent records for this species in the area. Recorded previously by Linton and Townsend as rare, but abundant in ditches north of Christchurch and around Sopley.
1 grid square;19

*Elodea canadensis Michx.
Canadian Waterweed

Introduced, frequent; streams and rivers. Recorded in the Stour, Avon, Moors River and the River Mude. Also planted in the Visitor Centre pond at Avon Forest Park in 1991. Introduced to England in 1842 (CTM), recorded at Radipole, Dorset in 1848 (Good, 1984). Described by Linton and Townsend as abundant in the Stour district.

*Elodea nutallii (Planchon)St.John
Nuttall's Waterweed

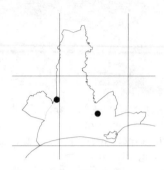

Introduced; slow-flowing streams and rivers. Two records for this rapidly spreading aquatic plant, one near the River Avon, the other in the River Stour. Also recorded in 1991 in the Moors River (Holmes, 1992). Certainly under-recorded and probably common throughout the Avon and Stour river systems.

JUNCAGINACEAE

Triglochin palustris L.
Marsh Arrowgrass

Uncommon; in marshes and very wet meadows. Several sites, all in marshes and very wet meadows by the River Avon. Previously recorded more frequently by Linton and Townsend, the decline is presumably due to drainage of suitable habitats.

Triglochin maritima L.
Sea Arrowgrass

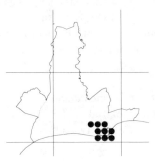

Infrequent; saltmarshes and muddy shores. Many records in saltmarsh turf around Christchurch Harbour. In Dorset and Hants almost confined to estuaries and harbours and not elsewhere along the coast.

POTAMOGETONACEAE

Potamogeton natans L.
Broad-leaved Pondweed

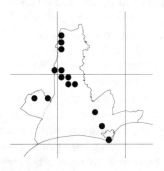

Infrequent; rivers, streams and nearby ditches. Recorded in the Moors River, and in ponds and ditches by the Stour and Avon Rivers.

Potamogeton polygonifolius Pourret
Bog Pondweed

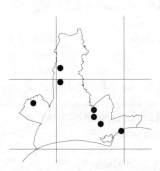

Infrequent; ponds, bogs and ditches. Recorded in ditches and boggy places in several parts of the area. Probably under-recorded. Recorded by Linton in many locations on heaths around Bournemouth which have since been swallowed up by the spread of Bournemouth.

Potamogeton lucens L.
Shining Pondweed

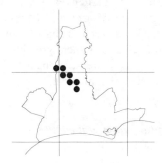

Infrequent; rivers and streams. Recorded only in the Moors River. Previously recorded as common by Linton and Townsend in the Stour and Avon Rivers. Probably under-recorded now, but also probably declined, mainly due to increased pollution of the rivers.

Potamogeton perfoliatus L.
Perfoliate Pondweed
Uncommon; rivers and streams. A single record in the River Avon, upstream from Knapp Mill in 1982. Recorded by Linton and Townsend from both the Stour and Avon Rivers. 1 grid square;19

Potamogeton crispus L.
Curled Pondweed
Uncommon; streams, ponds and slow-flowing rivers. One record in 1984 upstream from Knapp Mill in the River Avon. Also recorded in the Moors River in 1991 (Holmes, 1992). Recorded by Linton and Townsend also in the Stour and River Mude, much declined since then. 1 grid square;19

Potamogeton pectinatus L.
Fennel Pondweed

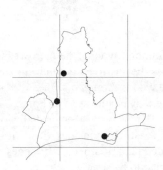

Uncommon; ponds, rivers and ditches. Present in the Moors River and also in the River Stour at Muscliff. Recorded by Linton and Townsend only from sites in the Moors River and River Stour.

Groenlandia densa (L.)Fourr.
Opposite-leaved Pondweed

Rare; clear nutrient-rich streams and ponds, usually markedly calcareous. Declining everywhere. Both records in the River Stour at Muscliff, in 1989 and 1991, although not in very large amounts. Recorded by Linton and Townsend as common, and now possibly under-recorded but more likely much decreased, due to pollution.

RUPPIACEAE

Ruppia maritima L.
Beaked Tasselweed
Rare; in brackish ditches and saltmarsh pools. Recorded in several pools on Stanpit Marsh in 1985. Previously recorded also near Hengistbury Head, Mudeford and at Wick in 1886 (Linton). Another decreasing species, and in the rest of Dorset recorded since 1980 only around Poole Harbour and the Fleet. 1 grid square;19

LILIACEAE

Narthecium ossifragum (L.)Hudson
Bog Asphodel

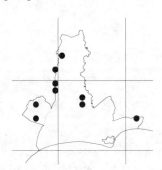

Infrequent; wet heaths and peaty bogs. Several records from boggy places near the Moors River, and on Kinson, Turbary and Chewton Commons. Not recorded from any heaths around

Avon Forest Park. Recorded by Linton as common in the Bournemouth area, but not now so common.

Kniphofia uvaria (L.)Oken
Red-hot-poker
Introduced, uncommon; an escape. One clump growing on the sandy cliff top west of Steamer Point woodland, Christchurch.
1 grid square;19

Convallaria majalis L.
Lily-of-the-valley

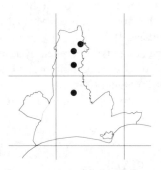

Infrequent as a native, but often a garden escape. The colony near Hurn in woodland might be native, the others are certainly garden escapes or relics of cultivation.

Polygonatum x hybridum Brugger
Garden Solomon's-seal
Uncommon; a garden escape, sometimes naturalized. A single record in 1991 from a hedgerow in Walkford, probably from dumped garden material.
1 grid square;29

Ruscus aculeatus L.
Butcher's-broom

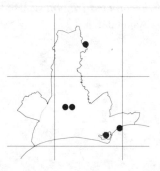

Infrequent; woods and hedges, but often planted. Several sites, all of which were almost certainly planted; two of them at least, for game cover.

Lilium spp.
Lily
Introduced; garden escapes. Two records of thriving plants by the edge of woodland, one near Hurn, the other at Matchams Lane, growing in rough grassy areas, probably from dumped garden material.
2 grid squares;19

Ornithogalum umbellatum L.
Star-of-Bethlehem

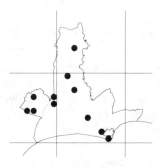

Infrequent; woods and grassy places, often planted, sometimes naturalized. Four sites on rough grassland by the River Stour, two in meadows by the Moors River, and one at Hengistbury Head. The Avon Forest Park site is almost certainly from a garden 'outcast', but the others appear to be naturalized. Not recorded in Linton or Townsend for this area.

Hyacinthoides non-scripta
(L.) sens.lat. Chouard ex Rothm.
Bluebell

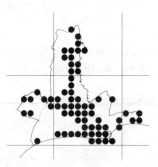

Common; woods, hedgerows and grassy heaths. Recorded in hedgerows and woods throughout the area, and on grassy heaths at Avon Forest Park and Hengistbury Head, and all along the coast. Linton described it as scarce near Bournemouth, it does not appear to be so now.

Hyacinthoides hispanica
(Miller)Rothm.
Spanish Bluebell
Introduced, infrequent; garden escape, sometimes naturalized. A colony in

woodland at Sopley Common, some at Wick by the river, and on Boscombe cliffs. Probably garden escapes and possibly hybrids with *H. non-scripta*. Undoubtedly under-recorded throughout the area.
3 grid squares;19

Muscari armeniacum
Leichtlin ex Baker
Garden Grape Hyacinth

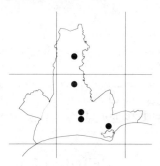

Introduced, infrequent; garden escapes on waste ground and roadside verges. Several sites on roadside verges and by car parks with plants which are obviously garden escapes, or possibly from dumped material.

Muscari comosum (L.)Miller
Tassel Hyacinth
Introduced, uncommon; sometimes planted and occasionally naturalized. Five plants were recorded flowering in 1985 in Avon Forest Park, but one or two of these were subsequently removed, and the others destroyed by mowing. No plants have been found since. There was a record of a plant in Dorset at Langton Matravers in 1929 (Good, 1948), and seen again in 1990.
1 grid square;10

Allium roseum L.
Rosy Garlic
Introduced, uncommon; garden escape. Four plants growing on the edge of an old rubbish tip overlooking Priory Marsh, Stanpit in 1990. Recorded in six other sites in Dorset since 1980.
1 grid square;19

Allium ursinum L.
Ramsons
Uncommon, prefers heavy soils. Alum Chine only, possibly originally planted or a garden escape, but growing well. Described by Linton as rare and not recorded in this area. Townsend lists only one site in the area, in Chewton Glen in private

grounds, where it may possibly still occur today.
1 grid square;09

***Allium triquetrum** L.*
Three-cornered Leek
Introduced, uncommon; hedgebanks and waste places in southwest England. Many plants in Middle Chine, some in Durley Chine and a few on the cliffs towards Alum Chine and at the Poole Boundary. Presumably planted, probably in Middle Chine; not mentioned in Linton or Townsend.
1 grid square;09

***Allium vineale** L.*
Wild Onion

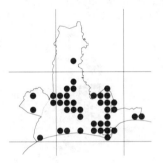

Introduced, fairly common; hedgerows and grassland. Recorded in the Stour valley, on roadsides to the east of the River Avon and along the coast. Recorded in the Stour valley and around Christchurch and Mudeford by Linton and Townsend. Few records in the north of the area.

AGAVACEAE

***Yucca filamentosa** L.*
Spanish-dagger

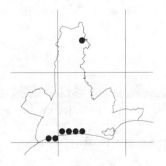

Introduced, infrequent; sometimes planted in gardens, occasionally an escape or relic of cultivation. Several records on the cliffs, probably planted but appearing naturalized; and one record in Avon Forest Park, previously a garden site.

***Cordyline australis** (Forst.)Endl.*
Cabbage-palm
Introduced, uncommon; sometimes planted near the coast. Recorded from the cliffs to the east of Boscombe Pier.
1 grid square;19

JUNCACEAE

***Juncus maritimus** Lam.*
Sea Rush

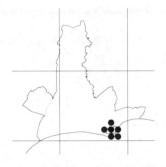

Infrequent; saltmarshes. Recorded from several saltmarsh sites around Christchurch Harbour.

***Juncus inflexus** L.*
Hard Rush

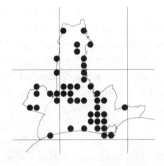

Fairly common; damp pastures and wet places, preferring heavy basic soils. Recorded in the Avon and Stour valleys, and elsewhere in wet flushes, especially along the cliffs.

***Juncus effusus** L.*
Soft Rush

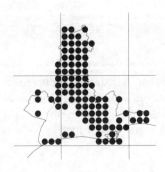

Very common; wet grassland, ditches and boggy places. Common in most places, except on the drier arable land to the east of Burton.

***Juncus conglomeratus** L.*
Compact Rush

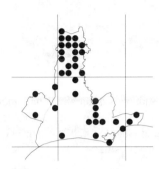

Frequent; wet places, especially on acid soils. Recorded mainly in the northern part of the area on peaty and sandy soils, and in the Avon valley.

***Juncus squarrosus** L.*
Heath Rush

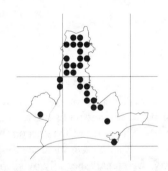

Frequent; sandy heathlands. Recorded on most of the sandy heaths, especially those in the north of the area.

***Juncus compressus** Jacq.*
Round-fruited Rush
Uncommon; wet alluvial meadows. One specimen from meadows north of Christchurch in 1983. Confirmation is needed as no fruits were present, and it can only be accepted at present as a 'possible' record. Not recorded in Dorset since the 1970's. Not recorded by Linton or Townsend for this area.
1 grid square;19

Juncus gerardi Loisel
Saltmarsh Rush

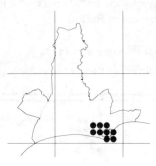

Infrequent; saltmarshes and muddy shores. Common on saltmarshes around Christchurch Harbour and in the upper estuary.

*Juncus tenuis Willd.
Slender Rush

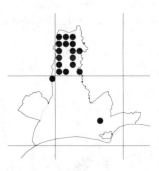

Introduced, infrequent; waste ground, and by tracks and paths. All records are from the north of the area on the drier sandy soils. On some sites it is fairly frequent, especially by sandy tracks. First recorded in Britain in 1883 and not mentioned in Linton or Townsend. First Dorset record apparently at Parkstone waterworks in 1921 (H.Bowen, pers.comm.). Now naturalized in many places and increasing its range steadily.

Juncus bufonius L. sens.lat.
Toad Rush

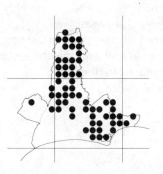

Common; wet grassy and muddy places, on damp tracks and paths and by ditches. Recorded widely throughout the area.

Juncus subnodulosus Schrank
Blunt-flowered Rush
Rare; fens, marshes and wet places with basic ground water. A single record in 1983 from Wick Meadows. No local sites are mentioned in Townsend or Linton but there are several other Dorset sites.
1 grid square;19

Juncus bulbosus L.
Bulbous Rush

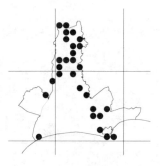

Frequent; wet and boggy places on heaths. Recorded mainly in the north of the area, in damp places on sandy heaths.

Juncus kochii F.W.Schultz
Infrequent; wet boggy heaths and meadows. Three sites, all in marshy meadows, mainly in the Avon valley. Previously common and now probably under-recorded but it is generally not considered to be a distinct species or subspecies now, but included in *J.bulbosus*.
3 grid squares;10,19

Juncus acutiflorus Ehrh.ex Hoffm.
Sharp-flowered Rush

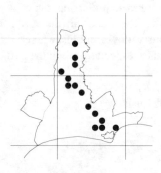

Infrequent; boggy meadows and wet heaths. Recorded from several scattered sites in boggy meadows in the Avon and Moors River valleys and by the River Mude.

Juncus articulatus L.
Jointed Rush

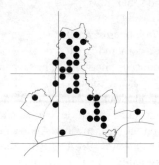

Frequent; wet meadows, ditches and streamsides. Common in wet meadows in the Avon and Moors River valleys, and wet places elsewhere.

Luzula campestris (L.)DC.
Field Wood-rush

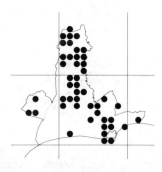

Fairly common; meadows and dry grassland. Recorded from many sites throughout the area, especially on dry heathy grassland.

Luzula multiflora (Retz.)Lej.
Heath Wood-rush

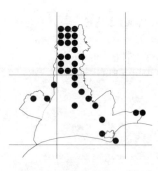

Frequent; heaths and woods on acid soils. Mainly recorded on the sandy soils in the north of the area, and on some grassy heaths elsewhere.

*Luzula luzuloides
(Lam.)Dandy & Wilmott
White Wood-rush
Introduced, uncommon; damp woods on acid soils. A single record in Ensbury Wood, Bournemouth, a damp oak woodland with heathy

clearings. Not mentioned in Linton or Townsend, not previously recorded in Dorset (VC9).
1 grid square;09

AMARYLLIDACEAE

Leucojum aestivum L.
Summer Snowflake
Very rare nationally; wet meadows and willow thickets; also cultivated and sometimes an escape or planted. Many plants in damp muddy patches, between a garden and the river at Wick, probably planted; and three clumps by a wet wood in the Stour valley near Hurn, perhaps native. Recorded by Good (1948) as a native in the Stour valley above Wimborne.
2 grid squares;19

Galanthus nivalis L.
Snowdrop

Introduced, infrequent; damp woods and by streams. Several records in damp woodland, probably all planted, but now well naturalized.

Narcissus spp.
Daffodil

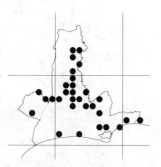

Frequent; damp grassland and woods. Many different species and hybrids grown in gardens, often these escape or are outcast, also sometimes planted in woods and other places, and become naturalized.

IRIDACEAE

Iris foetidissima L.
Stinking Iris

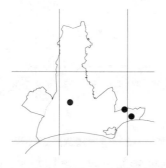

Infrequent; woods and copses. Often originally planted but sometimes native. These three sites possibly all planted. Described as 'abundant locally' in Linton, although only one site mentioned in this area.

Iris pseudacorus L.
Yellow Iris

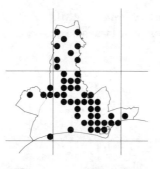

Fairly common; marshes, riversides, ditches and wet places. Common in the Avon, Stour and Moors River valleys. Also recorded elsewhere in wet places, including two records from wet flushes on the cliffs near Bournemouth.

Crocus vernus (L.)Hill sens.lat.
Spring Crocus

Introduced, infrequent; sometimes naturalized. Present on the Bournemouth cliffs, also four sites where it appears to be naturalized in woodlands, although the colonies are not extensive.

Crocus flavus Weston
Yellow Crocus
Introduced; naturalized in some places. Found on the cliffs at Fisherman's Walk, Bournemouth, where it may be naturalized or 'gardened'; and also in Avon Forest Park North in short dry grassland, probably a relic from a previous garden (pre-1976).
2 grid squares;10,19

Tritonia crocosmiflora (Lemoine)Nicholson
Montbretia

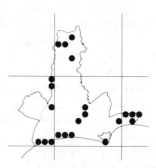

Introduced, frequent; a garden escape and often naturalized. Recorded mainly near the coast where it may be naturalized, but also scattered records elsewhere, usually as a garden escape.

Gladiolus communis L.
Eastern Gladiolus
Introduced, uncommon; sometimes naturalized in waste ground and on cliffs near the sea. Several plants growing on East Cliff, Bournemouth in 1990.
1 grid square;09

Tamus communis L.
Black Bryony

Infrequent; woods and hedgerows. Most records in hedgerows to the east of the River Avon, very few elsewhere. Only mentioned in a few locations in Linton and Townsend, although they describe it as frequent and common. Surprisingly infrequent, and perhaps needs searching for extremely thoroughly to confirm its continuing existence in most areas.

ORCHIDACEAE

Epipactis helleborine (L.)Crantz
Broad-leaved Helleborine
Uncommon; woods, clearings and shady banks. Four sites recorded, three with several plants present. All sites are on woodland edges, by tracks or paths, one site had many flowers present in 1985. Only one location mentioned in Linton and Townsend. Colonies seem to vary in numbers of plants from year to year. This is mainly due to changes in the weather, and because of these natural fluctuations they are more vulnerable to deer-browsing and other threats. Previously recorded in Chewton Glen in 1900 (Linton, 1919).
4 grid squares;10,19,29

Epipactis phyllanthes G.E.Sm.*var vectensis* (T & T.A.Steph.)D.P.Young
Pendulous-flowered Helleborine
Rare nationally, mainly in central southern England; woods and plantations, often on the margins. Two sites locally; one with a few plants struggling to survive under a hedgerow near Christchurch and unfortunately declining gradually since 1983; the other on the edge of woodland near Hurn, found in 1985. In one part of this site they appear to be declining, but in another part increased in numbers in 1991. However, in addition to being susceptible to dry weather in spring and early summer which can cause considerable fluctuations in annual numbers, they are also continually at risk of destruction by disturbance, sitework etc. and are therefore extremely vulnerable. Only three other sites in the rest of Dorset (Jenkinson).
2 grid squares;19

Spiranthes spiralis (L.)Chevall
Autumn Lady's-tresses

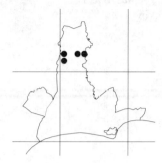

Uncommon; dry sandy or calcareous pastures. Four sites, all on short, dry heathy grassland, either mown or rabbit-grazed, two sites with large colonies of plants. The number of flower spikes present fluctuates from year to year influenced by the varying weather conditions. Recorded by Linton and Townsend as common at Mudeford in 1874, but listed in no other local sites then.

Listera ovata (L.)R.Br.
Common Twayblade

Uncommon; woods and shady places, also on grassy heaths. The sites recorded show a wide variety of habitats including in deciduous woodland, on a damp roadside verge, in grassy heathland under bracken and also growing under pine trees on acid heath. All colonies had several plants present. Previously recorded by Linton and Townsend also in meadows round Christchurch, but no recent records in these habitats.

Ophrys apifera Hudson
Bee Orchid
Uncommon; usually on dry calcareous grassland. A single plant recorded in Avon Forest Park in 1992, growing at the side of a gravel track across heathland, possibly with some calcareous hard-core beneath. Not recorded previously in this area (Linton, Townsend, Jenkinson).
1 grid square;10

Orchis morio L.
Green-winged Orchid
Infrequent; old unimproved meadows and dry grassland. Once common, but now a nationally threatened species. Sites are mainly in the north of the area and almost all are on dry heathy grassland in patches of open grass and amongst heather. Several colonies are quite large, but numbers seem to fluctuate from year to year; very recently numbers seem to have decreased on almost all sites. The nationally important site at St Leonards is under threat of development. Previously recorded by Linton and Townsend as common in the Stour watershed and around Herne.
8 grid squares;10,19

Dactylorhiza incarnata (L.)Soo
Early Marsh-orchid
Uncommon; very wet meadows with a high mineral content. Two colonies on wet heaths and two good colonies in floodmeadows to the east of the River Avon. The latter are of *subsp. incarnata* which is very local in Dorset. This species was known from here in 1893, but although it has survived well in places, it has declined and is now not at all widespread and still very vulnerable, especially to changes in water-table levels and grazing etc.
4 grid squares;19

Orchis morio

Dactylorhiza maculata subsp. ericetorum
(EFLinton)PFHunt&Summ
Heath Spotted-orchid
Infrequent; damp heaths and heathy grassland. Several sites, all on heaths or damp heathy grassland, some with colonies of many thousand plants. At Purewell Meadows several thousands of plants were destroyed by house building in 1989. Previously in 1983, one field was ploughed, then left to recover, and amazingly many orchids and other species did survive. Although frequent in the New Forest, and less so in the Poole Basin, this species is not common in the rest of Dorset. Unfortunately many sites in southeast Dorset, especially near St Leonards, Grange Estate and around Christchurch, are under constant threat of development.
12 grid squares;10,19,29

Dactylorhiza fuchsii (Druce)Soo
Common Spotted-orchid
Infrequent; damp meadows, marshes and grassland, usually on base-rich soils. Several sites, mainly in the north of the area. The habitats vary from roadside verges to grassy heathland, and one site is under pine trees; mostly there are only a few plants but in some places many plants are present, sometimes scattered over a wide area. On the site by a riverside walk, the single flowering spike was picked in 1991. There are almost 200 recorded sites in Dorset for this species, the majority being on or near the chalk and calcareous soils in north and west Dorset, and Purbeck. In this area, the southeast corner of the county, however, it is infrequent.
9 grid squares;09,10,19

Dactylorhiza majalis subsp. praetermissa
(Druce)DMMoore & Soo
Southern Marsh-orchid
Infrequent; wet meadows, marshes and marshy heathlands. Sites in 13 grid squares, mainly in the north of the area, and usually in flood meadows where the colonies are sometimes extensive. However, several sites are in marshy areas on heathland, and on marshy roadside verges. Linton recorded it 'about the Stour' but these sites seem to have gone; and the 33" tall plants in marshy ground near St Catherine's Hill seen in 1983, have also been destroyed. This species is found scattered widely across Dorset, but mainly in Purbeck and the west. The sites in eastern Dorset are permanently at risk of destruction by drainage, pollution and development.
13 grid squares;10,19

Dactylorhiza majalis subsp. praetermissa var junialis
'Leopard' Marsh-orchid
Uncommon; wet meadows and marshes. Two records for this variety of the Southern Marsh Orchid with ringed spots on the leaves. Both sites are near colonies of Southern Marsh Orchid, one in flood meadows of the River Avon, the other in damp heathy grassland.
2 grid squares;10,19

Dactylorhiza majalis subsp. traunsteinerioides (Pugsley) Bateman & Denholm
Narrow-leaved Marsh-orchid
Very rare; few sites nationally. Two sites in heathy grassland, one with a small population of var. bowmanii (Jenkinson). The St Leonards site is at risk from possible development. The nearest other sites are in Oxfordshire, North Hampshire and on the eastern side of the New Forest.
2 grid squares;10

Hybrid Marsh Orchids
Several hybrid Marsh Orchids have been recorded in the area, including hybrids of Southern Marsh Orchids with Common Spotted, Heath Spotted and also Early Marsh Orchids. These hybrids have all occurred with other populations of orchid species, either in Avon Forest Park or in the Avon valley north of Christchurch.
5 grid squares;10,19

Anacamptis pyramidalis
(L.)L.C.M.Richard
Pyramidal Orchid

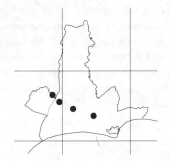

Uncommon; grassland and banks, on calcareous soils. A single site found in 1986, with three plants noticed. In 1988 over 60 were recorded on the same site, which was under threat from development. Plants were transplanted in 1988 and 1992 to three locations in the Stour valley in the hope of saving the species in this area. Many plants still left at the original site, which is to be developed soon. Not recorded in Linton or Townsend for this area.

PALMACEAE

*Trachycarpus fortunei
(Hook.)Wendl.fil.
Chusan Palm
Introduced, uncommon; on cliffs by the sea. One record from Highcliffe Castle grounds planted many years ago.
1 grid square;29

ARACEAE

Arum maculatum L.
Lords-and-Ladies, Cuckoo-pint

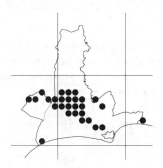

Frequent; woods and shady hedgebanks. Almost all records from hedgerows and thickets in the Stour valley. Records in other places on base-rich substrata would be expected. Described by Linton and Townsend as very common.

LEMNACEAE

Lemna polyrhiza L.
Greater Duckweed

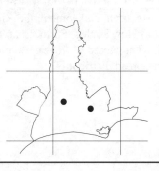

Uncommon; in still ditches and ponds. Two colonies recorded in 1984, one in ditches of meadows near the River Avon, and one in a backwater at Throop Mill by the River Stour. Very few previously recorded local sites, only in the Stour and Avon valleys.

Lemna trisulca L.
Ivy-leaved Duckweed
Uncommon; pools and ditches. A single record in 1982 in ditches near the River Avon north of Christchurch. Linton and Townsend recorded it as locally abundant in the Avon valley, now much more scarce.
1 grid square; 19

Lemna minor L.
Common Duckweed

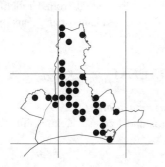

Frequent; ponds, ditches and still water. Recorded in ditches and still places in the Avon, Stour and Moors River valleys.

Lemna minuscula Herter
Least Duckweed

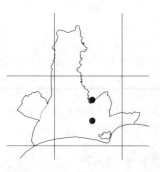

Introduced, rare; rapidly increasing in southern and eastern England. One colony recorded in 1991 in the River Avon just south of the railway line at Christchurch, and some in the River Avon below Sopley. Further searches will undoubtedly reveal a more widespread distribution.

Lemna gibba L.
Fat Duckweed
Uncommon; in still waters. A single record in 1984 in ditches in the Avon meadows north of Christchurch. Also recorded in the Moors River in 1991 (Holmes, 1992). Old records are almost all from the Avon valley; described as rare then, it is even more scarce now.
1 grid square; 19

SPARGANACEAE

Sparganium erectum L.
Branched Bur-reed

Frequent; slow-flowing rivers, streams, ditches and ponds. Recorded from the Avon, Stour and Moors Rivers, and the River Mude. More common in the smaller rivers.

Sparganium emersum Rehmann
Unbranched Bur-reed

Infrequent; shallow rivers, streams and ponds, but not in very acid water. Many records in the Moors River, one in the River Stour and two in the River Avon. Probably under-recorded in the Stour and Avon, although likely to be more frequent in the Moors River than the deeper rivers. Recorded as common in both the Stour and Avon valleys by Linton and Townsend.

TYPHACEAE

Typha latifolia L.
Bulrush

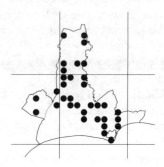

Frequent; ponds, ditches and riversides. Recorded in the Stour and Moors Rivers, also in the Avon valley and in the River Mude. Also present in ponds and ditches on heathlands.

Typha angustifolia L.
Lesser Bulrush

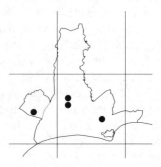

Rare; slow-flowing rivers, ditches and pools. Two sites in the River Stour and two in ponds, one fed by ditches from the River Avon. An additional site in a wet area just over the boundary in VC9. Previously recorded by Linton and Townsend in the Avon above Christchurch, not elsewhere in the area.

CYPERACEAE

Eriophorum angustifolium
Honckeny
Common Cottongrass

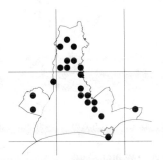

Frequent; wet heaths and bogs. Recorded from wet heaths and wet marshy meadows in the Avon valley, and on boggy heaths elsewhere.

Eriophorum vaginatum L.
Hare's-tail Cottongrass

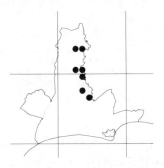

Rare; still decreasing in southern England. Wet heaths and bogs. Found scattered on wet heathlands with pools between the Moors River and the River Avon. Not common in any of the sites. Previously recorded by Linton and Townsend on heaths in the Avon valley and at Hengistbury Head.

Trichophorum cespitosum
(L.)Hartman
Deergrass

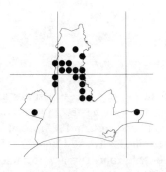

Frequent; heaths and damp peaty places. Mostly recorded on heaths to the west and north of Hurn, and around St Leonards.

Eleocharis parvula (Roem.& Schult.)Link
Dwarf Spike-rush
Very rare nationally; wet muddy places near the sea. One site locally but with only one reasonable sized colony, thus always at risk from being wiped out accidentally. The only post 1940 site in Dorset. A Red Data Book species, this has declined nationally from ten 10km squares to six since 1960. Previously found near Mudeford. Mansel-Pleydell described it at Little Sea as "abundant, and no chance of being extirpated by greedy collectors". It has not been seen there, nor at Redhorn Bay or Ower since 1936, but can easily be overlooked, although Good could find no sign of it in Little Sea in 1953 (Good, 1955).
1 grid square

Eleocharis multicaulis (Sm.)Sm.
Many-stalked Spike-rush
Uncommon; wet peaty places and sandy heaths. One colony recorded on the dried-up bed of a pond, on a heath north of Bournemouth. Linton and Townsend both described it as common on the heaths in this area.
1 grid square;19

Eleocharis palustris (L.)Roemer & Schultes
Common Spike-rush

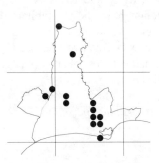

Infrequent; marshes, ditches and margins of ponds. Recorded mainly in the Avon valley and also by the River Stour.

Eleocharis uniglumis (Link)Schultes
Slender Spike-rush
Uncommon; open marshes near the coast. Recorded on Stanpit Marsh, from 1983 to present. Several scattered plants found in one area of the marsh. Recorded from only a few other sites in Dorset and Hampshire. Recorded by Rayner at Christchurch, but no definite records in Linton or Townsend for this area.
1 grid square;19

Scirpus maritimus L.
Sea Club-rush

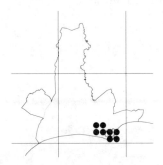

Infrequent; brackish pools, marshes and estuaries. Recorded in several places around Christchurch Harbour, forming fairly large stands in places on Stanpit Marsh.

Schoenoplectus lacustris (L.)Palla subsp.lacustris
Common Club-rush

Frequent; rivers, streams and ponds. Most commonly recorded in the Stour and Moors River valleys, and two records from the Avon valley. Not recorded also by Linton or Townsend in the Avon above Christchurch.

Schoenoplectus lacustris subsp.tabernaemontani
(C.C.Gmelin)A & D.Love
Grey Club-rush

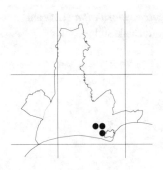

Uncommon; brackish water near the sea. Several sites around Christchurch Harbour. Although often frequent where it is found, there are not many recorded sites in Dorset.

Isolepis setacea (L.)R.Br.
Bristle Club-rush

Infrequent; damp sandy and gravelly places, and marshy meadows. Several sites, all in wet sandy meadows or heathy areas. Less sites than previously, due probably to loss or drainage of suitable habitats.

Eleogiton fluitans (L.)Link
Floating Club-rush

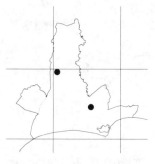

Uncommon; ditches, streams and ponds, especially on boggy heaths. One colony in ditches on marshes near the River Avon, the other on a dried-up pond bed on a heathy common. Townsend and Linton found it common all around Christchurch and Bournemouth. Probably now under-recorded, but also likely to have declined considerably.

Rhynchospora alba (L.)Vahl.
White Beak-sedge

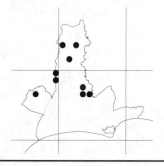

Infrequent; damp peaty places on heaths. Recorded from several scattered heathland sites, but also not recorded from many heathland areas. Described as fairly common by Linton, but much declined now, due to the loss of heathland.

Rhynchospora fusca (L.)Aiton fil.
Brown Beak-sedge
Rare nationally; wet heaths and bogs. A site with several plants by the edge of a pool in a wet heath near Hurn, and one on Week Common. Linton described it as locally frequent and often abundant, which it no longer is. Dorset has over half of the present UK population of this species (Mahon & Pearman, 1993).
2 grid squares;19

Carex paniculata L.
Greater Tussock-sedge

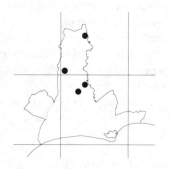

Infrequent; wet boggy woods and marshes. Four sites, all on peat, by the Moors River and River Avon. Not uncommon in these habitats, but suitable sites are few now. Previously common or frequent, the decline is due mainly to loss of and drainage of habitats.

Carex otrubae Podp.
False Fox-sedge

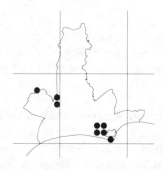

Infrequent; damp meadows and grassy places. Recorded in several places in Christchurch Harbour, and in the Stour valley. Described by Linton and Townsend as common, but not common now.

Carex divulsa subsp. divulsa Stokes
Grey Sedge

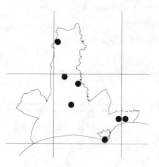

Infrequent; hedgerows, roadsides and grassland. A few scattered records in hedgerows and shady places. Perhaps under-recorded, as described by Linton and Townsend as common.

Carex spicata Hudson
Spiked Sedge

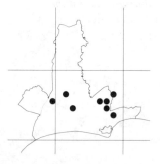

Infrequent; grassland and hedgebanks, on base-rich soils. Recorded in the Stour valley, and in hedgerows to the east of Burton. Not recorded by Linton or Townsend for this area.

Carex muricata subsp. lamprocarpa
Celak
Small-fruited Prickly-sedge

Infrequent; dry grassy pastures and hedgebanks. One record in Avon Forest Park North in 1985; specimen determined by A. Chater. Also recorded from Turbary Common, Hengistbury Head, and two records in 1987 near Mudeford. Probably under-recorded, previously much more frequent.

Carex disticha Hudson
Brown Sedge

Uncommon; damp meadows, fens and marshes. Several sites in three grid squares, all in wet meadows by the River Avon. Previously recorded by Linton and Townsend from the Avon and Stour valleys.

Carex arenaria L.
Sand Sedge

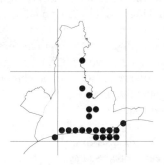

Frequent; sandy places, usually near the sea. Recorded on sandy cliffs along the coast, and in sandy places in the Avon valley. Recorded also in the Stour watershed by Townsend.

Carex remota L.
Remote Sedge

Frequent; hedgebanks, woods and damp shady places. Recorded throughout the area, except in the northern-most part, and most frequent to the east of Christchurch.

Carex ovalis Good.
Oval Sedge

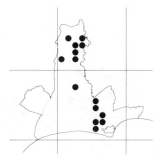

Infrequent; heaths and sandy grassland. Recorded from several sites in the Avon valley, and in and around Avon Forest Park. Recorded by Linton also in the Stour and Moors River valleys.

Carex echinata Murray
Star Sedge

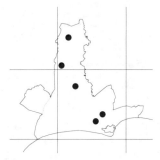

Infrequent; damp meadows and boggy places. A few sites in the Avon and Moors River valleys. Described in Linton and Townsend as very common, and occurring throughout the district. A species which has noticeably declined in frequency due mainly to loss and drainage of habitats.

Carex elongata L.
Elongated Sedge
Very rare nationally; one of the largest UK populations is in this area. Two records, one in 1985, the other in 1990 from sites within a mile of each other, both near the Moors River. Not mentioned for this area in any of the old Floras. Declining nationally due to loss and drainage of habitats.
2 grid squares;10

Carex curta Good.
White Sedge
Rare; bogs, acid fens and marshes. Two sites near the Moors River, one found in 1982, the second seen in 1990. Also one plant found in woods

near the River Stour in 1989. Few other sites in Dorset. Recorded in several other sites in the Avon and Moors River valleys by Linton and Townsend, so probably declined now. 3 grid squares;09,10

Carex hirta L.
Hairy Sedge

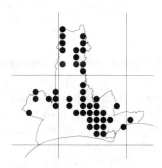

Fairly common; rough grassland, damp meadows and woods. Recorded mainly in and around the Avon valley, and in the Stour valley. Also in Avon Forest Park and nearby.

Carex acutiformis Ehrh.
Lesser Pond-sedge

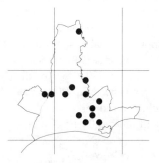

Infrequent; stream and riversides and wet meadows. Recorded in the Avon and Stour valleys, and by the Moors River.

Carex riparia Curtis
Great Pond-sedge

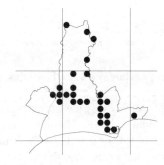

Frequent; riversides, streams and ditches. Mainly recorded in the Avon valley, also in the Stour and Moors River valleys.

Carex pseudocyperus L.
Cyperus Sedge

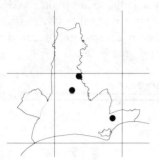

Rare; ditches and wet woods. One site by the Moors River in a much overgrown meadow in 1983, and one site in very wet carr, on peat, found in 1989. A third record near ponds in Mudeford Wood in 1987 may have been introduced during or after re-shaping of the ponds. Previously recorded at several sites in the Moors River and Avon valleys by Linton and Townsend.

Carex rostrata Stokes
Bottle Sedge
Uncommon; wet peaty bogs and ditches. One site near the Moors River, and one in meadows near the River Avon. Both sites are very wet, although the latter site is at risk from drying out gradually. One additional site just in VC9. Not nearly as common now as in 1900, and likely to decline in future.
3 grid squares;09,10,19

Carex vesicaria L.
Bladder-sedge

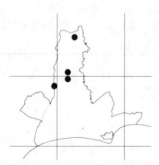

Rare; wet boggy places. Four sites, all in wet areas, in woodland edges or scrub. Mostly very few plants but one site with a good colony of large plants. One site is just in VC9. Much more common previously according to Linton, who described it as "plentiful along the Stour and Moors Rivers"

Carex pendula Hudson
Pendulous Sedge

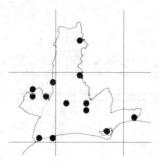

Infrequent; damp shady riverbanks and woods. Recorded from the Stour and Avon valleys, also on East Cliff and at Durley Chine, Bournemouth and Chewton Bunny. Recorded by Linton and Townsend from banks of the Stour.

Carex sylvatica Hudson
Wood-sedge

Uncommon; woods and hedgerows. Three sites, all in the far east of the county, in dense woodland. Very few other previous records for this species in this area, but generally common throughout Britain.

Carex flacca Schreber
Glaucous Sedge

Infrequent; grassland and marshes. Recorded from several areas of heathy grassland, mostly with soil of a high calcareous content. Probably much under-recorded, previously listed by Linton and Townsend as common, although only a few sites mentioned.

Carex panicea L.
Carnation Sedge

Infrequent; wet heaths, fens and damp grassy places. Recorded mainly in wet meadows in the Avon valley, and scattered records elsewhere. Probably under-recorded, described by Linton and Townsend as common.

Carex distans L.
Distant Sedge
Uncommon; marshes near the sea. A single record, in 1991, on a marshy site on which it had previously been recorded, but not found recently. Described as common and frequent in Townsend and Linton, but not, apparently, now.
1 grid square;19

Carex binervis Sm.
Green-ribbed Sedge
Uncommon; heaths and rough grassland. A single record from Kinson Common in 1992. A few sites listed in Linton and Townsend for this area, but not many. Probably declined due mainly to loss of the heaths around Bournemouth.
1 grid square;09

Carex demissa Hornem.
Common Yellow-sedge

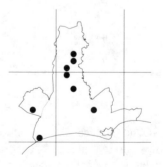

Infrequent; damp meadows and boggy places. Recorded in several places in the Moors River valley and in Avon Forest Park.

Carex caryophyllea Latourr.
Spring-sedge
Uncommon; dry, usually calcareous grassland. One record from Hengistbury Head in 1993, growing amongst heathy grass at the edge of woodland. Previously recorded by Townsend as common, and by Linton as frequent, although only listed for Winton.
1 grid square;19

Carex pilulifera L.
Pill Sedge

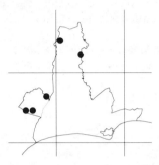

Infrequent; grassland, heaths and woodland on sandy or peaty acid soils. One site found in 1986 in the Stour valley in damp woodland with heathy patches. Also recorded on Turbary Common, and near Ashley. Previously described by Linton and Townsend as common around Bournemouth and in the Avon watershed; not so common now.

Carex acuta L.
Slender Tufted-sedge

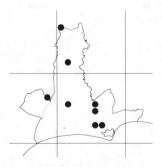

Infrequent; riverbanks, streams and marshes. Recorded in several places by the River Avon, also by the Stour and Moors River. Few records in the west of Dorset and in southwest England.

Carex nigra (L.)Reichard
Common Sedge

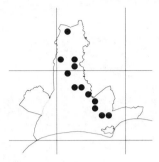

Infrequent; wet grassy places and boggy heaths. Several sites in wet meadows in the Avon valley, and on heaths between the Rivers Stour and Avon.

GRAMINAE

Festuca pratensis Hudson
Meadow Fescue

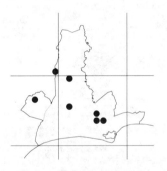

Infrequent; wet grassy places, especially old meadows. Recorded in meadows in the Avon and Stour valleys, and by the Moors River.

Festuca arundinacea Schreber
Tall Fescue

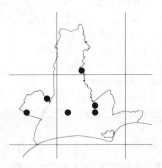

Infrequent; damp grassy meadows. Five sites in little or un-grazed meadows by rivers, three by the Avon and two near the Stour; also at Turbary Common. Not recorded in this area by Linton or Townsend, but this was probably as much due to taxonomic differences as to its infrequency. Probably still under-recorded.

Festuca gigantea (L.)Vill.
Giant Fescue

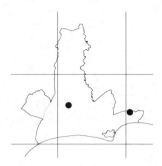

Uncommon; moist woods and shady hedgebanks. One site on the edge of woodland near Hinton Admiral; the other between Throop and Hurn, not far from the site mentioned in Linton and Townsend.

Festuca rubra L. agg.
Red Fescue

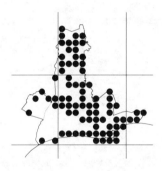

Very common; grassland, waste ground and sandy places. Recorded widely throughout the area, especially in the river valleys and along the coast.

Festuca juncifolia St-Amans
Rush-leaved Fescue

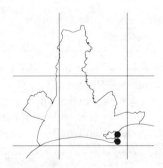

Uncommon; sand dunes. Two unconfirmed records in 1981 from Hengistbury Head, one incomplete specimen "could probably" be this species (pers.comm. C.Stace). Other specimens from the site have since been collected by P.Bowman and determined by C.Stace as this species.

Festuca tenuifolia Sibth.
Fine-leaved Sheep's-fescue

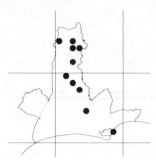

Infrequent; heaths and sandy grassland. Recorded from a few scattered sites, mainly on heaths, probably much under-recorded; some records included in *Festuca ovina* agg. Recorded as abundant by Townsend on East Bournemouth Cliffs and in the Stour watershed.

Festuca ovina L. agg.
Sheep's-fescue

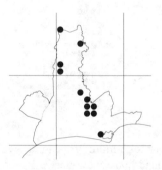

Infrequent; dry grassland. Recorded mainly on and around Town Common. Recorded by Linton as frequent, but only listed for Turbary Common and Mudeford.

*Festuca brevipila Tracey
Hard Fescue
Introduced, uncommon; often sown in grass-seed mixtures, naturalized on roadsides. Recorded in 1993 on the grassy verges of the slip road from Verwood onto the A31. Probably widespread and under-recorded.
1 grid square;10

x Festulolium loliacium (Festuca pratensis x Lolium perenne)
Hybrid Fescue

Uncommon; damp meadows. Recorded from wet meadows in the Avon valley, and by the River Stour at Throop. Few previous records for this area, and only occasionally in Dorset. Recorded from Mudeford and Christchurch by Townsend, and at Burton (Linton, 1925).

Lolium perenne L.
Perennial Rye-grass

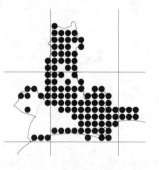

Very common everywhere; grassland and waste places. Recorded in almost all grid squares, possibly absent in a few mainly heathland squares.

*Lolium multiflorum Lam.
Italian Rye-grass
Introduced, uncommon; often cultivated and naturalized. A single record at Muscliff, Bournemouth in grassy river meadows.
1 grid square;09

Vulpia fasciculata (Forskal)Samp.
Dune Fescue
Rare; dunes and sandy ground. A single site with many plants, seen in 1993 on sandy ground at Hengistbury Head. Three sites recorded in Dorset since 1960, and few in southwest England. Not previously recorded in old Floras for this area.
1 grid square;19

Vulpia bromoides (L.)S.F.Gray
Squirreltail Fescue

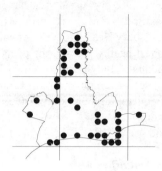

Frequent; dry sandy and gravelly places. Many records along the coast, and on dry sandy grassland and heaths.

Vulpia myuros (L.)C.C.Gmelin
Rat's-tail Fescue

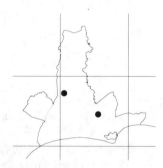

Uncommon; dry sandy places and walls. One record at East Parley on a grassy roadside verge, and one on a spoil heap from excavations. A few records in Linton and Townsend for this area, not many for the rest of Dorset.

Desmazeria rigida (L.)Tutin
Fern-grass

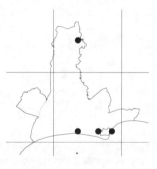

Uncommon; dry banks and walls, mainly in calcareous areas. One record in 1986 on dry grassy sand and shingle near Spellars Point on Stanpit Marsh, and one in 1992 at Mudeford. Also recorded on Boscombe Cliffs and at Ashley Heath.

Desmazeria marina (L.)Druce
Sea Fern-grass

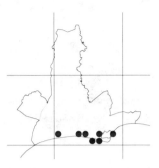

Infrequent; dry sandy places near the sea. Recorded in sandy places on cliffs and walls from Bournemouth to Mudeford. Described as not common and rare by Linton and Townsend, and only listed for Mudeford.

Poa annua L.
Annual Meadow-grass

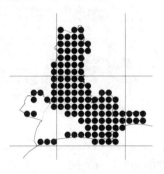

Very common everywhere; grassland, cultivated and waste ground. Recorded in all but three grid squares, almost certainly present in all squares.

Poa infirma Kunth
Early Meadow-grass
Extremely rare nationally; short dry sandy grassland near the sea. One site on Hengistbury Head recorded in 1987 and refound in 1993. Also recorded in 1993 on cliffs at Boscombe. Both sites with a few tiny plants, with anthers about 0.2-0.3mm long. Previously only known from the Channel and Scilly Isles, Cornwall and Devon; this site is by far the furthest east (pers.comm. D.Pearman). First collected in Britain on the Lizard in 1876, and not recorded again until 1950 (Hubbard). Depauperate examples of *Poa annua* can easily be confused with this species.
2 grid squares,19

Poa bulbosa L.
Bulbous Meadow-grass
Very rare nationally; dry sandy areas near the coast, in southern Britain. Three surviving colonies growing around Christchurch Harbour all in trampled eroded sandy areas. A species always at risk, as too much erosion will destroy its habitat, and with not enough trampling it will be ousted by other species. Only two other Dorset sites, none in Hants. Not recorded previously in old Floras for this area.
3 grid squares;19

Poa nemoralis L.
Wood Meadow-grass

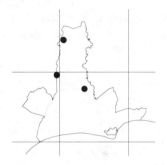

Uncommon; woods and shady places. One record from near the Avon Causeway Road under trees in 1986, another from a shady track at Parley in 1987, just in VC9; also at Ashley in 1993. Recorded as very rare by Linton, and there are also few records from Dorset.

Poa pratensis sens.lat.
Smooth Meadow-grass

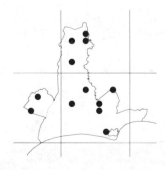

Infrequent; meadows and grassy places. Some widely scattered records, undoubtedly much under-recorded.

Poa trivialis L.
Rough Meadow-grass

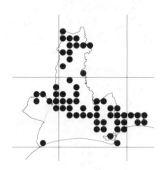

Common; grassland, hedgerows and shady places. Recorded mainly in the Avon, Stour and Moors River valleys, and in hedgerows and woods in the east of the area.

Puccinellia maritima (Hudson)Parl
Common Saltmarsh-grass

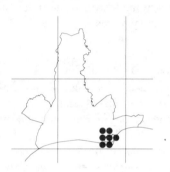

Infrequent; saltmarshes and muddy estuaries. All sites recorded are in and around Christchurch Harbour, where it is locally common. Locally abundant in the few saltmarshes in Dorset.

Puccinellia distans (Jacq.)Parl
Reflexed Saltmarsh-grass
Uncommon; saltmarshes and sandy places near the sea. A single record from Stanpit Marsh in 1985. Recorded previously from here by Salter (Linton, 1925). Few records in Dorset.
1 grid square;19

Dactylis glomerata L.
Cock's-foot

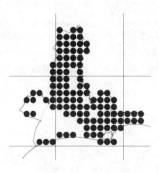

Very common everywhere; grassland, woods, hedgerows and disturbed ground. Recorded from almost all grid squares, and probably present in all squares.

Cynosurus cristatus L.
Crested Dog's-tail

Frequent; grassland and roadsides. Recorded mainly in the Avon valley, also by the Stour and other rivers, and in grassy places elsewhere.

Cynosurus echinatus L.
Rough Dog's-tail
Introduced; casual growing in waste and sandy places near the coast. A single colony on Bournemouth Cliffs about 3m x 1m in size, found in 1990. It remains to be seen whether this species survives here, or lives up to its 'casual' status.
1 grid square; 09

Catabrosa aquatica (L.)Beauv.
Whorl-grass

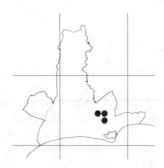

Uncommon, and apparently decreasing in abundance; ditches and shallow streams. Two sites, both in ditches in meadows on the edge of the River Avon floodplain. Described by Linton as common about Christchurch, and throughout the Avon valley. Certainly not as frequent now.

Briza maxima L.
Great Quaking-grass

Introduced, uncommon; a casual or garden escape. Three records; two on Bournemouth East Cliff, and one in Highcliffe, on a tarmac path west from Nea Meadows, by a block of flats. The latter is either an escape, or possibly grown from seed dropped accidentally; while the former may have been planted.

Glyceria fluitans (L.)R.Br.
Floating Sweet-grass

Frequent; ditches, pools and very wet meadows. Recorded in the Avon, Stour and Moors River valleys and by the River Mude.

Glyceria plicata Fr.
Plicate Sweet-grass

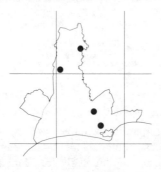

Uncommon; streams and ditches. Found in sites varying from wet marshy meadows to pools on heathy commons. Described by Linton as frequent in wet places but cannot now be regarded as such in this area, even allowing for some under-recording.

Glyceria maxima (Hartm.)Holmberg
Reed Sweet-grass

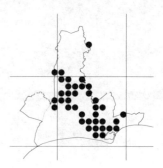

Frequent; rivers, streams and ditches. Recorded in the Avon, Stour and Moors River valleys and by the River Mude.

Bromus sterilis L.
Barren Brome

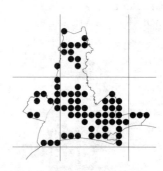

Common; roadsides, dry grassy places and disturbed ground. Most frequent to the east of the River Avon, along the coast and in the Stour valley. Fewer records in the north of the area.

Bromus erectus Hudson
Upright Brome

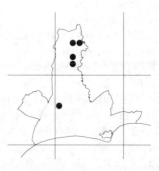

Infrequent; dry grassy banks and downs, usually calcareous, but also on sandy soils. Two sites in Avon Forest Park in dry sandy grassland, also at Ashley and Throop. Not common in this area or in the New Forest, but frequent on the chalk and limestone in Dorset.

Bromus hordeaceus L. sens.lat.
Soft-Brome

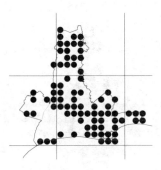

Very common; grassland, waste places and disturbed ground. Recorded mainly in the Stour and Avon valleys, and along the coast.

Bromus commutatus Schrader
Meadow Brome
Uncommon; previously common in old meadows. One site in Winkton Meadows north of Christchurch. Possibly overlooked and under-recorded, as Linton described it as common around Bournemouth. However, only one dot is shown in this area in the Atlas of the British Flora.
1 grid square;19

Brachypodium sylvaticum
(Hudson)Beauv.
False Brome

Frequent; woods and shady hedgebanks. Records widely scattered, although most common in the eastern part of the area.

Leymus arenarius (L.)Hochst.
Lyme-grass

Infrequent; sandy cliffs and seashores. Recorded in many places along the coast. One record by the River Stour on a sandy bank, almost certainly planted, previously to 1988. Recorded by Linton as rare and local, from North Haven to Highcliff, no other locations given. Described by Townsend as very rare, between Bournemouth and Highcliff 1886-1889, and no other Hampshire sites listed. Recorded at Southbourne (Linton, 1925). Apart from Poole Harbour and Studland, very few sites in the rest of Dorset.

Elymus caninus (L.)L.
Bearded Couch

Uncommon; hedgebanks, fields and woods. Two records, both in grassy places by the River Stour, at Muscliff and Throop.

Elymus repens (L.)Gould
Common Couch

Fairly common; fields, waste places and disturbed ground. Mainly recorded in the Stour valley, and to the east of Burton, very few records in the north of the area.

Elymus pycnanthus
(Godron)Melderis
Sea Couch

Infrequent; dunes and saltmarshes on coasts and in estuaries. Several sites around Christchurch Harbour, and on the cliffs. Found in Dorset around Poole Harbour and in the Weymouth area, and in Hampshire around Lymington and the Solent.

Elymus farctus (Viv.)Runemark ex Melderis
Sand Couch
Uncommon; sandy shores, especially young dunes. Recorded in 1993 on the new dunes at Hengistbury Head. Previously recorded by Linton and Townsend at Hengistbury, Mudeford and on Bournemouth Cliffs.
1 grid square;19

Hordeum secalinum Schreber
Meadow Barley

Infrequent; meadows and grassy places. Several records in the Stour valley, and one in the Avon valley. Listed in Townsend and Linton for this area, only in Stour meadows and from Christchurch to Mudeford.

Hordeum murinum L.
Wall Barley

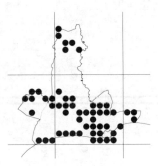

Fairly common; waysides, waste places and dry places, often by walls. Recorded all along the coast, in the Stour valley and east of Burton. Only a few records in the north of the area.

*Hordeum vulgare L.
Six-rowed Barley
Introduced; a relic of cultivation. Several plants growing on a roadside verge in front of a hedgerow in Riverside Lane, Bournemouth, in 1986.
1 grid square;19

*Avena fatua L.
Wild-oat
Introduced, uncommon; cultivated fields and waste ground. One record in Avon Forest Park in 1985 in a disturbed area of ground by Matchams Lane.
1 grid square;10

Arrhenatherum elatius (L.)Beauv.ex J&CPresl.
False Oat-grass

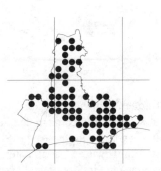

Common; grassy places, especially roadside verges. Recorded throughout the area, but more frequently in the Stour and Avon valleys, and to the east of the River Avon.

Trisetum flavescens (L.)Beauv.
Yellow Oat-grass
Uncommon; dry grassland especially on calcareous soils. One record from a grassy verge near Throop, Bournemouth. Described by Linton and Townsend as common in the Stour and Avon watersheds.
1 grid square;19

*Lagurus ovatus L.
Hare's-tail
Introduced, perhaps native in the Channel Isles, uncommon; sometimes naturalized in sandy places. Recorded on a grassy slope on the cliffs at Southbourne, probably introduced. Not mentioned in Linton or Townsend, although recorded as a casual in the locality before 1962 (ABF).
1 grid square;19

Deschampsia cespitosa (L.)Beauv.
subsp. cespitosa
Tufted Hair-grass

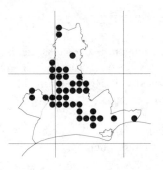

Fairly common; ditches, wet meadows and woods. Recorded mainly in the Stour and Moors River valleys, and in parts of the Avon valley.

Deschampsia flexuosa (L.)Trin.
Wavy Hair-grass

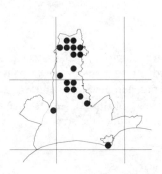

Infrequent; dry grassy heaths. Several records from heaths between the Moors River and the River Avon and around Ashley, also at Muscliff and Hengistbury Head.

Aira praecox L.
Early Hair-grass

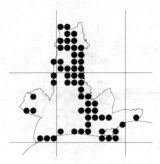

Fairly common; dry sandy grassland, heaths and banks, and other dry sandy places. Recorded on almost all dry sandy heathlands, and along the coast.

Aira caryophyllea L.
Silver Hair-grass

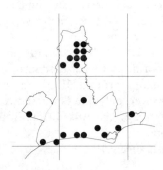

Frequent; dry sandy and gravelly places. Several records along the coast, and also in and around Avon Forest Park. Much less frequent than *Aira praecox*.

Anthoxanthum odoratum L.
Sweet Vernal-grass

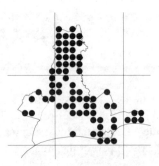

Common; grassland, roadsides and heaths. Recorded widely throughout the area, especially around St Leonards. Not common along the coast.

Holcus lanatus L.
Yorkshire-fog

Very common everywhere; fields, woods, grassy and waste places. Recorded in almost all grid squares, probably present in all squares.

Holcus mollis L.
Creeping Soft-grass

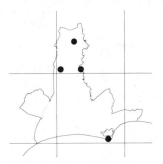

Uncommon; edges of woods and shady places. Several scattered records, all on the edges of woodland. Undoubtedly under-recorded, although much less frequent than *H.lanatus.*

Agrostis curtisii Kerguelen
Bristle Bent

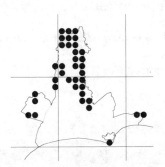

Frequent; dry sandy heaths. Mainly recorded in the northern part of the area, where it is sometimes common; also at Kinson, Turbary and Chewton Commons and Hengistbury Head. Recorded previously by Townsend also on the cliffs between Bournemouth and Boscombe.

Agrostis canina L. sens.lat.
Brown Bent

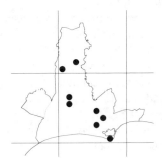

Infrequent; damp heaths and acid grassland. A few widely scattered records. Recorded by Linton and Townsend as common; and by Good (1948) as "distribution uncertain in detail". Apparently not as common as previously thought.

Agrostis capillaris L.
Common Bent

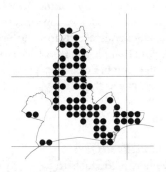

Common; heaths, roadsides and acid grassland. Recorded throughout the area, surprisingly few records on Bournemouth cliffs.

Agrostis gigantea Roth
Black Bent

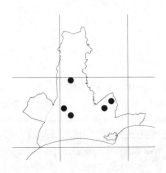

Infrequent; arable land, fields and grassy places. Two records in the Stour valley, one in the Moors River valley and two in the arable area to the east of Burton. Not recorded by Linton and Townsend in this area.

Agrostis stolonifera L.
Creeping Bent

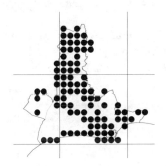

Very common; damp grassy places. Recorded throughout the area but noticeably less common on the drier arable areas to the east of the River Avon.

Ammophila arenaria (L.)Link
Marram

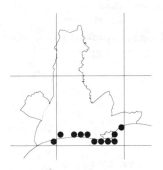

Infrequent; dunes and sandy places. Recorded in many places along the coast, including on cliff tops. Recorded by Linton and Townsend as locally abundant from Bournemouth to Mudeford.

Calamagrostis epigegos (L.)Roth
Wood Small-reed
Uncommon; ditches, fens and damp woods. One record from Avon Forest Park, on heathland, in a damp ditch. Not recorded previously in old Floras for this area.
1 grid square;10

Phleum pratense L. **subsp. pratense**
Timothy

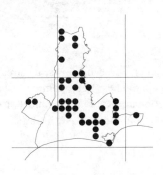

Frequent; grassland and roadsides. Recorded mainly in the Stour and Avon valleys, and scattered elsewhere.

Phleum pratense subsp. bertolonii (DC.)Bornm.
Smaller Cat's-tail

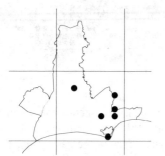

Infrequent; dry grassland. A few scattered records, mainly in the east of the area. Previous distribution is very uncertain.

Alopecurus pratensis L.
Meadow Foxtail

Frequent; damp meadows and roadsides. Recorded mainly in the Stour and Moors River valleys, and by the River Mude.

Alopecurus geniculatus L.
Marsh Foxtail

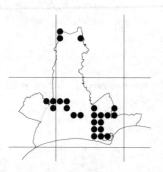

Frequent; wet meadows, by streams and in ditches. Very common in the Avon valley, and also recorded in the Stour and Moors River valleys.

Alopecurus x plettkei Mattfeld *(Alopecurus bulbosus x A. geniculatus)*
Very rare; saltmarshes. A single site at Stanpit Marsh seen in 1991. First found in Britain in 1977, and on Stanpit Marsh in 1980. Very few sites known in Britain.
1 grid square;19

Alopecurus bulbosus Gouan
Bulbous Foxtail

Rare nationally and locally; saltmarshes and meadows near the coast. Several records from different areas of Stanpit and Priory Marshes, but the number of plants present is not known as this is very difficult to determine. No other local sites, very rare.

Parapholis strigosa (Dumort.)C.E.Hubbard
Hard-grass

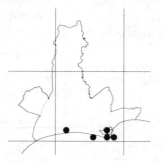

Infrequent; saltmarshes and sandy places near the sea. Five sites around Christchurch Harbour, three of these on saltmarsh, one from sandy grassland and one site on sandy cliff tops at Boscombe. Recorded as common by Linton and Townsend, but only listed by them for this area at Mudeford.

Phalaris arundinacea L.
Reed Canary-grass

Fairly common; streamsides, ditches and wet places. Recorded in the Stour, Avon and Moors River valleys, by the River Mude and also in large ditches elsewhere.

*Phalaris canariensis L.
Canary-grass

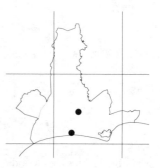

Introduced, uncommon; casual in waste places. A record from Riverside Lane, Iford in 1986, and also recorded on Bournemouth Cliffs. Previously recorded around Bournemouth, and as well-established in the sand at Mudeford (Linton, Townsend). Also recorded on Stanpit Tip in 1956 (BNSS).

Phragmites australis (Cav.)Trin.ex Steudel
Common Reed

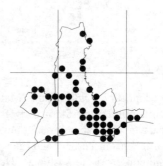

Fairly common; riverbanks, streams and ditches. Recorded mainly in the lower Avon and Stour Rivers and Christchurch Harbour, in the Moors River and also in wet flushes along the cliffs.

*Cortaderia selloana
(Schult.)Ashers.& Graebn.
Pampas-grass

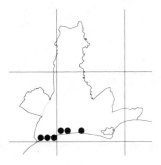

Introduced, infrequent; a garden escape. Many plants on the cliffs between Alum Chine and Southbourne, probably originally planted and persisting well.

*Pseudosasa japonica (Sieb & Zucc.ex Steudel)Makino ex Nakai
Bamboo

Introduced, infrequent; a garden escape. Several sites along the cliffs, and a few scattered additional records elsewhere.

Danthonia decumbens(L.)DC.
Heath-grass

Infrequent; heaths and acid grassland. Four records in damp, heathy, acid grassland in the Avon valley, and one record on Chewton Common. Described as common by Linton and Townsend, apparently much declined now.

Molinia caerulea (L.)Moench
Purple Moor-grass

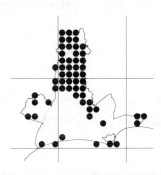

Fairly common; wet heaths and boggy places. Recorded in damp places on most heaths, especially common in the north of the area. Also recorded in several places in wet flushes on the cliffs.

Nardus stricta L.
Mat-grass
Uncommon; heathland and peaty soils. A small amount on Coward's Marsh, Christchurch, recorded in 1984. Recorded by Linton as locallly abundant around Bournemouth and in the Stour watershed, and surprisingly no other recent local records. A considerably declined species.
1 grid square;19

Cynodon dactylon (L.)Pers.
Bermuda-grass
Very rare; sandy shores in southern England. A colony growing in sandy ground in a chine near Bournemouth, seen in 1993, and on Bournemouth East Cliff. Previously recorded in Dorset between Studland and Branksome (Good, 1984) and by Poole Harbour (Linton). Also recorded for this area of VC11 at Bournemouth (Rayner).
1 grid square;09

Spartina anglica C.E.Hubbard
Common Cord-grass

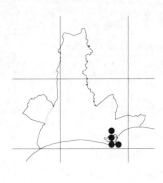

Infrequent; saltmarshes and mud-flats. Many small clumps growing in several sites around Christchurch Harbour. First recorded at Lymington in 1892, by 1907 it had spread to East Dorset, and by 1960 was widespread around British coasts. Subsequently generally declining, but locally gradually spreading, at least on Stanpit Marsh, although only very slowly.

*Digitaria sanguinalis (L.)Scop.
Hairy Finger-grass
Introduced, rare; waste land, railway sidings and sandy arable fields. Several dozen plants seen in 1993 growing at two sites at the side of pavements in Pokesdown. The location is just within the urban area excluded from this Flora, but it has been described as, in future, this plant may continue to spread locally. The number of plants has apparently increased at both sites since it was first reported in 1990. Previously recorded as a casual in Poole (Good, 1984).
1 grid square;19

Cynodon dactylon

113

APPENDIX 1.

Native species recorded historically in the area but not refound during the present survey (excluding the 40+ individual *Rubus* and *Rosa* species).

Lycopodium clavatum	Wimborne Heath	(Linton)
Oreopteris limbosperma	Highcliff 1879	(Townsend)
Asplenium ceterach	Christchurch 1920	(Rayner)
Pilularia globulifera	near Bournemouth	(Townsend)
Ranunculus sardous	Hengistbury	(Townsend)
	Throop, Holdenhurst	(Linton)
Ranunculus auricomus	Shrubbery, Bournemouth	(Townsend)
Ranunculus fluitans	Rivers Stour and Avon	(Townsend)
Ranunculus circinatus	Christchurch1893	(Townsend)
Papaver lecoqii	Christchurch 1888	(Townsend)
	Southbourne 1924	(Rayner)
Papaver argemone	Hengistbury, Kinson	(Townsend)
	Throop, Christchurch	(Linton)
Glaucium flavum	Hengistbury Head	(Townsend)
Fumaria bastardii	Bournemouth, Mudeford	(Townsend)
Lepidium ruderale	Christchurch	(Linton, 1919)
Erysimum cheiranthoides	Christchurch	(Rayner)
Descurainia sophia	Knapp Mill 1921	(Dorch.Mus.Herb)
Reseda lutea	Bournemouth, Mudeford	(Townsend)
Viola reichenbachiana	Bourneouth West 1889	(Townsend)
Silene noctiflora	Christchurch 1923	(Rayner)
Silene maritima	Mudeford	(Townsend)
Silene conica	Highcliff	(Rayner)
Cerastium arvense	Bournemouth, Christchurch	(Townsend)
Scleranthus annuus	Kinson, Highcliff, Stour watershed	(Linton)
Chenopodium polyspermum	Iford, Highcliff, Heron Court	(Townsend)
Chenopodium murale	Herne, Christchurch, Mudeford	(Townsend)
	Hengistbury 1920	(Rayner)
Chenopodium urbicum	East Cliff Bournemouth, Burton	(Townsend)
Chenopodium hybridum	Bournemouth	(Linton)
Atriplex glabriuscula	Hengistbury Head, Mudeford	(Townsend)
Geranium sanguineum	Christchurch	(Townsend)
Geranium columbinum	Bournemouth	(Townsend)
	Kinson, Highcliff	(Linton)
Geranium purpureum	Christchurch	(Townsend)
Vicia lutea	Knapp Mill 1924	(Rayner)
Medicago polymorpha	Christchurch	(Linton, 1925)
Trifolium scabrum	Hengistbury Head, Mudeford	(Townsend)
	Throop, Muscliff	(Linton)
Trifolium incarnatum	Throop, Christchurch Station	(Linton)
	Bournemouth	(Rayner)
Anthyllis vulneraria	Chewton Cliffs	(Townsend)
	Southbourne	(Linton, 1925)
Potentilla argentea	Hengisbury, Highcliff 1888	(Townsend)
	Ensbury, Wick, Herne	(Linton)
	Christchurch	(Rayner)
Alchemilla vulgaris	Avon valley 1879	(Townsend)
Prunus avium	Chewton Farm	(Linton)
Umbilicus rupestris	Iford	(Townsend)
Drosera anglica	Ramsdown	(Townsend)
Myriophyllum verticillatum	near Sopley	(Townsend)
Myriophyllum alterniflorum	Avon valley	(Townsend)
	St Leonards Bridge	(Linton)
Callitriche platycarpa	Bournemouth, Stour watershed	(Townsend)
Callitriche obtusangula	Iford, Sopley Common 1889	(Townsend)
Callitriche hamulata	Avon	(Linton)
Eryngium maritimum	Hengistbury, Mudeford	(Townsend)
Scandix pecten-veneris	Bournemouth	(Linton)
Pimpinella saxifraga	Chewton	(Linton)
Sium latifolium	Herne, Iford 1849	(Townsend)
	Tuckton	(Linton, 1925)
Oenanthe aquatica	Throop, Holdenhurst, Christchurch	(Townsend)
Silaum silaus	Stour valley	(Townsend)
	Christchurch, Chewton	(Linton)
Sison amomum	Herne Station	(Linton)
Torilis nodosa	Hengistbury	(Townsend)
	Christchurch, Highcliff	(Linton)
Euphorbia exigua	Kinson, Highcliff, Pokesdown	(Linton)
Euphorbia paralias	Hengistbury Head 1879, Mudeford 1879	(Townsend)
Polygonum mite	Kinson, Wick 189	(Linton)
	Christchurch 1925	(Rayner)
Polygonum minus	Christchurch, Sopley	(Linton)

Fallopia dumetorum	Jumpers Common 1888	(Townsend)
	Iford Bridge, Ensbury	(Linton)
Rumex pulcher	Bournemouth 1879, Mudeford, Herne	(Townsend)
	Burton	(Linton)
Rumex crispus x hydrolapathum	St Leonards 1958	(PBNSS 48,27)
Ulmus glabra	Christchurch	(Linton)
Salix aurita	Alum Chine 1889	(Townsend)
	Kinson	(Linton)
Vaccinium myrtillus	Iford 1924	(PDNHAS 46,194)
Limonium binervosum	Christchurch Harbour 1885	(Linton)
Anagallis minima	Chewton Common 1889	(Townsend)
Cicendia filiformis	Boscombe Chine 1863, Chewton Common 1889	(Townsend)
Gentianella amarella	Chewton Common 1889	(Townsend)
Cynoglossum officinale	Kinson	(Linton)
Lithospermum arvense	Bournemouth 1889	(Townsend)
Echium vulgare	Mudeford, Highcliff, Herne Station 1889	(Townsend)
	Kinson	(Linton)
Hyoscyamus niger	Mudeford	(Townsend)
	Christchurch	(Rayner)
Verbascum pulverulentum	Christchurch	(Rayner)
Kickxia spuria	Bournemouth, Hinton 1888	(Townsend)
Kickxia elatine	Pokesdown, Highcliff 1889, Chewton	(Townsend)
Euphrasia brevipila	Hengistbury Head	(Linton)
Euphrasia borealis	Herne 1895	(Townsend)
Orobanche rapum-genistae	Southbourne	(Linton, 1925)
Pinguicula vulgaris	Christchurch 1813	(Linton)
Utricularia minor	near Herne Station, Sopley, Heron Court	(Linton)
Verbena officinalis	Holdenhurst, Throop, Herne, Highcliff, Burton	(Linton)
Mentha pulegium	Christchurch	(Townsend)
Mentha x verticillata	Herne Bridge	(Linton)
Thymus pulegioides	Chewton Common, Highclif	(Linton)
Thymus praecox	Christchurch, Bournemouth	(Townsend)
Nepeta cataria	near Sopley	(Townsend)
Lobelia urens	Christchurch	(Linton, 1925)
Legousia hybrida	Moordown	(Linton)
Valerianella rimosa	Winton	(Linton)
Senecio erucifolius	Highcliff 1889, Chewton	(Townsend)
Petasites hybridus	Christchurch	(Linton)
Filago lutescens	Bournemouth to Christchurch, Pokesdown, Hengistbury	(Townsend)
	Herne 1894	(Linton)
	Herne 1922	(Linton, 1925)
Filago pyramidata	Stour watershed	(Townsend)
Gnaphalium sylvaticum	Peat Moors River 1898	(Linton)
	Highcliff	(Townsend)
Anthemis arvensis	Wick, Christchurch	(Linton)
Anthemis cotula	Kinson, Herne, Pokesdown, Throop	(Linton)
Chamaemelum nobile	Bournemouth, Christchurch	(Townsend)
	Holdenhurst, Chewton	(Linton)
Otanthus maritimus	Mudeford 1879	(Townsend)
Artemisia absinthium	Hengistbury 1903	(Townsend)
	Mudeford	(Linton, 1919)
Carlina vulgaris	Chewton Cliffs	(Townsend)
Carduus acanthoides	Herne Bridge	(Linton)
Cirsium acaule	north of Herne Station	(Linton)
Arnoseris minima	Pokesdown, Wick 1879, Iford 1892	(Linton)
	Moors River, Mudeford	(Linton, 1925)
Hypochoeris glabra	Hengistbury Head, Highcliff	(Townsend)
	Mudeford 1891	(Townsend)
Lactuca virosa	Bournemouth	(Rayner)
Baldellia ranunculoides	Herne, Christchurch, Chewton	(Linton)
Damasonium alisma	Christchurch	(Townsend)
Zostera marina	Bournemouth	(Townsend)
	Southbourne	(Linton, 1925)
Zostera noltii	Christchurch Harbour 1882	(Townsend)
	Stanpit Marsh	(Linton)
	Mudeford 1926	(Rayner)
Potamogeton x salicifolius	Christchurch	(Linton)
Potamogeton pusillus	Moors River, Highcliff, Avon	(Linton)
Zannichellia palustris	Mudeford 1888	(Linton)
Simethis planifolia	Bournemouth 1879	(Townsend)
	Mudeford 1915	(NCC)
Asparagus officinalis	Christchurch 1799, Highcliff 1902	(Townsend)
	Mudeford 1922	(Linton, 1925)
Juncus effusus x inflexus	Peat Moors River	(Linton)
	Southbourne 1926	(Rayner)
Sisyrinchium bermudiana	Christchurch	(Townsend)
Gladiolus illyricus	Ensbury	(Linton)
Epipactis palustris	Chewton Glen	(Linton)
Listera cordata	Bournemouth 1853	(Linton, 1919)

Neottia nidus-avis	Highcliff 1879		(Townsend)
Eriophorum gracile	Herne		(Linton)
Eleocharis acicularis	Christchurch		(Rayner)
Scirpus sylvaticus	Chewton Bunny 1893		(Townsend)
	Hurn 1903		(Linton, 1919)
	Hurn 1928		(Rayner)
Isolepis cernua	Hengistbury, Mudeford 1879, Highcliff 1888		(Townsend)
Schoenus nigricans	Kinson		(Linton)
Carex diandra	Herne Bridge 1879		(Townsend)
	Holdenhurst, Burton, Knapp Mill		(Linton)
Carex extensa	Hengistbury, Christchurch, Mudeford		(Townsend)
Carex hostiana	Herne Bridge		(Townsend)
Carex flava	Kinson		(Linton)
Carex serotina	Chewton Common, Herne, Ensbury		(Linton)
Carex pallescens	Herne		(Townsend)
Carex pulicaris	Herne Station, by the Stour, Burton		(Linton)
Puccinellia fasciculata	Hengistbury		(Townsend)
Briza minor	Boscombe, Christchurch		(Townsend)
Glyceria x pedicellata	Wick, Christchurch		(Townsend)
Glyceria declinata	Christchurch		(Townsend)
Bromus madritensis	Christchurch Harbour 1879		(Townsend)
	Knapp Mill 1923		(Rayner)
Bromus rigidus	Bournemouth		(Linton, 1919)
	Bournemouth cliffs in plenty 1922		(Rayner)
Bromus racemosus	Christchurch, Throop		(Linton)
Elymus farctus x pycnanthus	Bournemouth, Mudeford		(Townsend)
Deschampsia setacea	Ensbury		(Linton)
Gastridium ventricosum	Hengistbury, Highcliff 1888, Mudeford		(Townsend)
Polypogon monspeliensis	Knapp Mill 1923		(Rayner)

APPENDIX 2.

In addition to those seen and found during the present survey, the following native species have been recorded by other botanists in the area since 1980 (excluding *Rubus* and *Rosa* species).

Oreopteris limbosperma	St Leonards	1985	NCC
Ranunculus fluitans	Moors River	1991	N.Holmes
Glaucium flavum	Hengistbury Head	1989	R.Walls
Viola palustris	Hurn	1991	B.Edwards*
Lotus angustissimus	Wick	1992	D.A.Pearman*
Potentilla argentea	Hurn	1991	B.Edwards
Callitriche obtusangula	Moors River	1991	N.Holmes
Polygonum mite	Hurn	1991	B.Edwards
Monotropa hypopitys	Hurn Forest	1990	J.White*
Euphrasia micrantha x nemorosa	Avon Forest Park	1990	R.P.Bowman*
Utricularia minor	Week	1986	R.Walls
Thymus praecox	Hengistbury Head	1980-89	R.P.Bowman
Valerianella eriocarpa	Bournemouth	1993	R.Walls*
Alisma lanceolatum	Moors River	1991	N.Holmes*
Potamogeton berchtoldii	Moors River	1991	N.Holmes*
Zannichellia palustris	Stanpit Marsh	1982-86	R.Walls
Juncus ambiguus	Stanpit Marsh	1980-86	R.P.Bowman*
Scirpus sylvaticus	Iford	1992	B.Edwards
Carex x boenninghausiana	St Leonards	1982-87	R.W.David*

* denotes a species not previously recorded in this area.

BIBLIOGRAPHY

Aflalo, F.G. (ed.) (1905). <u>Half a century of sport inHampshire.</u> London, Country Life.

BEE (Bournemouth Evening Echo) (1993). <u>'A year of almost constant warmth'</u>. 9.1.93, p25.

Bristow, C.R., Freshney, E.C. & Penn, I.E. (1991). <u>Geology of the country around Bournemouth.</u> London, HMSO.

Bursche, E.M. (1971). <u>A handbook of water plants.</u> London, Frederick Warne.

Clapham, A.R., Tutin, T.G. & Moore, D.M. (1989). <u>Flora of the British Isles.</u> 3rd ed. Cambridge, Cambridge University Press.

Clapham, A.R., Tutin, T.G. & Warburg, E.F. (1962). <u>Flora of the British Isles.</u> 2nd ed. Cambridge, Cambridge University Press.

Dandy, J.E. (1958). <u>List of British vascular plants.</u> London, British Museum.

Dandy, J.E. (1969). <u>Watsonian vice-counties of Great Britain.</u> London, Ray Society.

Dony, J.G., Jury, S.L. & Perring, F.H. (1986). <u>English names of wild flowers.</u> 2nd ed. London, Botanical Society of the British Isles.

Driver (1816). Sale notice for Barnsfields. (Hampshire County Record Office).

Findlay, D.C. (1984). <u>Soils and their use in South West England.</u> Harpenden, Soil Survey of England and Wales.

Garrard, I. & Streeter, D. (1983). <u>The wild flowers of the British Isles.</u> London, Macmillan.

General Register Office. (1901). <u>Census of Population 1901: County Report Hampshire.</u> London, HMSO.

Good, R. (1948). <u>A geographical handbook of the Dorset flora.</u> Dorchester, Dorset Natural History & Archaeological Society.

Good, R. (1955). <u>First addendum to the hand-list of the Dorset flora.</u> Reprinted from the Proc. DNHAS. Vol 75, pp. 157-163. Dorchester, Dorset Natural History & Archaeological Society.

Good, R. (1970). <u>Hand-list of the Dorset flora (second addendum).</u> Reprinted from the Proc. DNHAS 83 (1961). Dorchester, Dorset Natural History & Archaeological Society.

Good, R. (1984). <u>A concise flora of Dorset.</u> Dorchester, Dorset Natural History & Archaeological Society.

Haslam, S.M., Sinker, C.A & Wolseley, P.A. (1982). <u>British water plants.</u> London, Field Studies Council.

Holmes, N.T. (1992). <u>Moors River SSSI. Draft management plan.</u> Unpublished report for English Nature.

Hubbard, C.E. (1984). <u>Grasses.</u> 3rd ed. (revised by Hubbard, J.C.E.) Harmondsworth, Penguin Books.

Jenkinson, M.J. (1991). <u>Wild orchids of Dorset.</u> Gillingham, Dorset, Orchid Sundries Ltd.

Jermy, A.C., Arnold, H.R., Farrell, L. & Perring, F.H. (1978). <u>Atlas of ferns of the British Isles.</u> London, Botanical Society of the British Isles.

Jermy, A.C. & Camus, J. (1991). <u>The illustrated field guide to ferns and allied plants of the British Isles.</u> London, Natural History Museum Publications.

Jermy, A.C., Chater, A.O. & David, R.W. (1982). <u>Sedges of the British Isles.</u> BSBI Handbook No. 1. London, Botanical Society of the British Isles.

Keble Martin, W. (1965). <u>The concise British flora in colour.</u> London, Ebury Press and Michael Joseph.

Kenchington, F.E. (1944). <u>The Commoners' New Forest.</u> London, Hutchinson.

King, M.P. (1974). <u>Beneath your feet. The geology and scenery of Bournemouth.</u> Swanage, Purbeck Press.

Linton, E.F. (1900). <u>Flora of Bournemouth.</u>

Linton, E.F. (1919). <u>Flora of Bournemouth Appendix I.</u>

Linton, E.F. (1925). <u>Flora of Bournemouth Appendix II.</u>

Lousley, J.E. & Kent, D.H. (1981). <u>Docks and knotweeds of the British Isles.</u> BSBI Handbook No. 3. London, Botanical Society of the British Isles.

Mabey, R. (1972). <u>Food for free.</u> London, Collins.

Mahon, A. & Pearman, D. (1993). <u>Endangered wildlife in Dorset. The county red data book.</u> Dorchester, Dorset Environmental Records Centre.

Mansel-Pleydell, J.C. (1874). <u>The flora of Dorsetshire.</u> 2nd ed. 1895. Dorchester.

Meikle, R.D. (1984). <u>Willows and poplars of Great Britain and Ireland.</u> BSBI Handbook No. 4. London, Botanical Society of the British Isles.

Melville, R.V. & Freshney, E.C. (1982). <u>British regional geology: The Hampshire Basin and adjoining areas.</u> 4th ed. London, HMSO.

Mitchell, A. (1978). <u>A field guide to the trees of Britain and Northern Europe.</u> 2nd ed. London, Collins.

Morris, Sir D. (1914). Woods and forests. In Morris, Sir D. (ed.) <u>A natural history of Bournemouth and district.</u> Bournemouth, Bournemouth Natural Science Society, pp182-9.

Nilsson, S. (1979). <u>Orchids of Northern Europe.</u> Harmondsworth, Penguin Books.

OPCS. (1991). <u>Census of Population 1991: County Report Dorset.</u> London, HMSO.

Pearman, D. (ed.) (1991-3). <u>Recording Dorset</u> No. 1-3. Dorchester, Dorset Environmental Records Centre.

Perring, F.H. (1978). <u>Critical supplement to the Atlas of the British flora.</u> 2nd ed. Wakefield, Botanical Society of the British Isles, EP Publishing.

Perring, F.H. & Farrell, L. (1983). <u>British red data books, 1. Vascular plants.</u> 2nd ed. Lincoln, Royal Society for Nature Conservation.

Perring, F.H. & Walters, S.M. (1982). <u>Atlas of the British flora.</u> 3rd ed. Wakefield, Botanical Society of the British Isles, EP Publishing.

Phillips, R. (1978). <u>Trees in Britain, Europe and North America.</u> London, Pan Books.

Proceedings of the Bournemouth Natural Science Society 1908-

Proceedings of the Dorset Natural History and Archaeological Society, Dorchester. 1974-

Rayner, J.F. (1929). <u>A supplement to Frederick Townsend's Flora of Hampshire and the Isle of Wight.</u> Swaythling, Southampton.

Rich, T.C.G. (1991). <u>Crucifers of Great Britain and Ireland.</u> BSBI Handbook No. 6. London, Botanical Society of the British Isles.

Rich, T.C.G. & Rich, M.D.B. (1988). <u>Plant crib.</u> London, Botanical Society of the British Isles.

Roles, S.J. (1957-65). <u>Flora of the British Isles illustrations.</u> Cambridge, Cambridge University Press.

Rose, F. (1981). <u>The wild flower key.</u> London, Frederick Warne.

Ross-craig, S. (1948-74). <u>Drawings of British plants.</u> Parts 1-31. London, Bell.

Stace, C.A. (1991). <u>New flora of the British Isles.</u> Cambridge, Cambridge University Press.

Townsend, F. (1904). <u>Flora of Hampshire including the Isle of Wight.</u> 2nd ed. London, L. Reeve.

Tutin, T.G. (1980). <u>Umbellifers of the British Isles.</u> BSBI Handbook No. 2. London, Botanical Society of the British Isles.

Walls, E. (1929). <u>The Salisbury Avon.</u> Bristol, Arrowsmith.

Wigginton, M.J. & Graham, G.G. (1981). <u>Guide to the identification of the more difficult vascular plant species.</u> Banbury, NCC England Field Unit Occasional Paper No. 1. Nature Conservancy Council.

Wilson, P.J. (1992). Britain's arable weeds. <u>British Wildlife.</u> 3 (3), 149-161.

INDEX OF ENGLISH AND LATIN GENERIC NAMES
Page numbers in bold type refer to colour plates and illustrations